Technological Change, the Learning Curve and Profitability

For Jennifer and Christopher

Technological Change, the Learning Curve and Profitability

Dudley Jackson

Edward Elgar
Cheltenham, UK • Northampton, MA, USA

Published by
Edward Elgar Publishing Limited
8 Lansdown Place
Cheltenham
Glos GL50 2HU
UK

Edward Elgar Publishing, Inc.
6 Market Street
Northampton
Massachusetts 01060
USA

A catalogue record for this book
is available from the British Library

Library of Congress Cataloguing in Publication Data

Jackson, Dudley.
Technological change, the learning curve and profitability /
Dudley Jackson.
Includes index.
1. Technological innovations—Economic aspects. 2. Learning curve
(Industrial engineering) I. Title.
HC79.T4J33 1998
338'.064—dc21 98–17083
CIP

ISBN 1 85898 881 0

Typeset by Robert Hood at the University of Wollongong

Printed and bound in Great Britain by MPG Books Ltd, Bodmin, Cornwall

Contents

List of Figures

List of Tables

PART I

TECHNOLOGICAL CHANGE

1 Introduction

1.1 Methodology and scope

This book explains the economic analysis of two distinct phenomena which are very important to the production of goods and services: technological change and learning. As an integral part of this explanation the book also provides 'real world' plant- or industry-level data and case studies to illustrate and exemplify the concepts and theories relating to technological change and learning. (The concept of technological change and the classification of the categories of technological change are explained in Chapter 2, and the concept of learning, as it applies to production, is explained in Chapter 6.)

In order to understand any phenomenon, theoretical concepts, classification schemes (explicating those concepts), and models (showing how the relevant concepts are related to each other) must be exemplified by 'real world' empirical data to illustrate the concepts and to show how the models actually work. Either without the other leads to only imperfect understanding: concepts, classification schemes and models without data are 'empty' and thus incapable of leading to the requisite *operational* understanding of the phenomenon, while any set of empirical data without organizing concepts, classification schemes and models is 'blind' because the forces behind the data cannot be seen and *analytically* understood.

Accordingly, the methodology of this book is consistently to ensure that all concepts and models are exemplified by 'real world' empirical data. On occasion, understanding 'real world' data requires an introductory step of using simplified invented 'hypothetical' data to elucidate the basic application of a concept and/or a model; where necessary this book makes use of such hypothetical data but always identifies these data as such.

With regard to the 'real world' data, it should be noted that plant-level data relating to technological change and learning are remarkably difficult to obtain, largely because such data contain, when current, highly sensitive commercial information, the confidentiality of which is closely protected. Subsequently when no longer current such data 'disappear' into company archives generally never to be seen again. However, a certain amount of data can be found, especially in the published studies of the United States Department of Labor Bureau of Labor Statistics, which has a long and honourable tradition of researching and reporting on matters relating to productivity. Accordingly, this book draws heavily on these studies, supplemented where possible by the research of others. This book can thus claim to be unique in providing 'real world' empirical illustrations of technological change and learning set in the context of a rigorous theoretical framework.

Technological change and learning reduce the cost of production and thereby increase profit and profitability (assuming that the price of the product is not decreased in a matching way – and it should be noted that, under the stimulus of competition to gain market share, the gains of technological change and learning are often 'shared' with users of the product through price reductions). Furthermore, as explained in Chapter 5, technological change often improves the quality of a product, also to the benefit of product users.

Technological change and learning reduce the cost of production by increasing the productivity of inputs, or, put the other way round, by reducing the quantity of inputs required per unit of output. The economic analysis of technological change and learning concentrates on this aspect of increased input productivity or, alternatively, decreased 'unit input requirements'.

Chapter 2 explains neutral technological change and exemplifies this with empirical data on the change in multifactor productivity in United States manufacturing industry over the period 1948 to 1981 – the calculation of multifactor productivity being fully explained in the chapter. Chapter 3 explains the important theory of non-neutral technological change, and Chapter 4 describes and analyses capital-using/labour-saving non-neutral technological change in the switch from steam locomotives to diesel–electric locomotives in the United States railroad industry over the period 1948 to 1967. Chapter 5 considers issues relating to unit total cost and quality under technological change and illustrates these issues with two case studies: one on technological change in a pulp-making plant; the other on technological change in petroleum refining to produce gasoline of improved quality over the period 1913 to 1955.

Chapter 6 explains how learning affects the production process, using as illustrations empirical real world statistics on an individual worker assembling propelling pencil mechanisms and statistics on the production of Liberty ships in shipyards in the United States during the Second World War. Chapter 7 uses both these sets of data to explain the proper derivation of the learning curve (explaining also the various variants of the learning curve) and concludes by explaining how the learning curve reduces unit cost. Using both sets of data again, Chapter 8 explains the time constant progress function, which is the function required in order to describe and analyse how the learning curve affects the time rate of production and so how the learning curve affects profit and the rate of profit.

The remainder of this introductory chapter explains some basic concepts of productivity, unit input requirement, unit cost, unit net margin, profit and the rate of profit.

1.2 Productivity and unit cost

Technological change and learning generally have as a main objective or result an improvement in the productivity of inputs and thence a reduction in cost. The first stage in analysing technological change and learning is to understand the relation between inputs and output and thence in the second stage to understand any change in the ratio between inputs and output, leading on to a third stage of understanding the resulting decrease in cost and increase in profitability.

Inputs to the process of production may be divided into three main categories: the inputs of materials and non-factor services; the factor input of labour; and the factor input of fixed capital. Chapter 5 considers the impact of technological change on the inputs of materials and the concluding part of Chapter 7 explains how learning may affect unit materials cost, so these matters will be left until then.

The factor input of labour can be measured relatively straightforwardly in terms of labour-hours per annum. The number of workers employed can also be used as a (relatively crude) measure of labour input, but the obvious problem is that if the average labour-hours worked per worker changes then the quantity of labour services provided is not correctly measured by the number of workers employed. Hence in most cases throughout this book the factor input of labour will be measured as a quantity of labour-hours, and this quantity is denoted by L.

Capital is a produced means of production. (By contrast, natural resources are a non-produced means of production.) There are two types of capital: fixed capital and circulating capital (alternatively known as inventories). Fixed capital comprises those items which have a repeated, on-going use in the process of production; that is, whenever an item of fixed capital is used in the process of producing a unit of output it remains available to be used in the process of producing the next unit of output. The opposite of fixed capital is circulating capital, where the defining characteristic of circulating capital is that it has a once-for-all use in the process of production; that is, whenever an item of circulating capital is used in the process of producing a unit of output it is 'used up' and is not then available for producing the next unit of output. To illustrate, the sewing machine of a dress-maker is fixed capital; the dress fabric or sewing cotton of a dress-maker is circulating capital.

The factor input of fixed capital is most appropriately measured, for purposes of analysing production, by the dollar amount that is spent on acquiring and installing the items of fixed capital which make up a plant (or any production facility). This has the corollary that the prices for all the items of fixed capital must be in terms of the prices of a constant base year (with the further corollary that all other prices must be adjusted to the same base year). This measure of fixed capital I call the 'acquisition cost' of fixed capital, and, measured at constant prices, the acquisition cost of fixed capital is denoted by K. The acquisition cost of fixed capital is, effectively, the gross final expenditure on fixed capital, but this final expenditure is then used as a measure of the capacity of the *stock* of fixed capital installed. (The rationale for measuring the quantity of fixed capital through the final expenditure on gross fixed capital formation, but taken as a stock installed (and known technically in the system of national accounts as the 'gross fixed capital stock'), is explained in Chapter 1 of my book *Profitability, Mechanization and Economies of Scale* (Ashgate, 1998).)

The quantity of output will generally be measured by the physical quantity of output produced per annum (or any other relevant production period, but annual data are mostly used throughout this book). In the case of the learning curve the quantity of output may be measured in batches. Note that whenever output is to be taken relative to the input of fixed capital, annual output, and only *annual* output, must be used. The reason for this is twofold. First, depreciation, which measures the resource cost of using fixed capital in the process of production, is nearly always calculated on an *annual* basis. Second, the rate of profit is, with very few exceptions, calculated as a rate *per annum* (that is, with reference to the annual flow of profit). Together, this means that all flows taken relative to the stock of fixed capital should be annual flows. The physical quantity of output is denoted throughout this book by Q.

Productivity may be understood in two ways, both of which will be used in this book. First, and most commonly, productivity may be measured as a ratio of output to input. Thus labour productivity may be measured as the ratio Q/L. For an hypothetical illustration, suppose a brick-making plant uses 500,000 labour-hours per annum, L, to produce 150,000,000 bricks per annum, Q, then labour productivity may be measured as:

Labour productivity	=	Q/L
	=	150,000,000 bricks per annum/500,000 labour-hours per annum
	=	300 bricks per labour-hour

However, there is another important way of measuring labour productivity, and this is the way which is mainly used in this book. We can consider productivity in terms of the quantity of input which is required per unit of output; this is, so to speak, an inverse measure of productivity – the smaller the measure, the greater the productivity – and this measure is known as *unit input requirement*. Unit input requirement is defined and measured as the quantity of input divided by the quantity of output. (It must be clearly understood that in all like terms using the word 'unit' – such as unit input requirement, unit cost, or unit net margin – the word 'unit' means *per unit of* ***output***.) For example, unit labour requirement is defined and measured as L/Q. Using the previous example:

Unit labour requirement	=	L/Q
	=	500,000 labour-hours per annum/150,000,000 bricks per annum
	=	0.003333 labour-hours per brick
	=	3.333 labour-hours per 1,000 bricks

The significance of this way of measuring labour productivity is that if we multiply unit labour requirement by the price of labour (or the 'wage rate per labour-hour'), the result is the cost of labour per unit of output, or 'unit labour cost'. To illustrate, suppose the price of labour, denoted by p_L, is $2.40 per labour-hour, then:

Unit labour cost	=	$p_L \times (L/Q)$
	=	$2.40 per labour-hour × (500,000 labour-hours per annum/150,000,000 bricks per annum)
	=	$2.40 per labour-hour × 0.003333 labour-hours per brick
	=	$0.008 per brick
	=	$8 per 1,000 bricks

Unit labour cost may be more simply understood as the total payment to labour (or the 'total annual wage bill'), $p_L L$, divided by the annual quantity of output, Q – which gives the cost of labour per unit of output – but it is analytically more useful to understand unit labour cost as the price of labour multiplied by unit labour requirement.

For the input of fixed capital, unit capital requirement is defined as the dollar value (at constant prices) of fixed capital installed (the 'acquisition cost') divided by the quantity of annual output, or K/Q. As already explained, it is most important that, with regard to fixed capital, only the *annual* quantity of output be used. To illustrate, suppose the brick-making plant had an acquisition cost of fixed capital installed of \$2,000,000, K, then:

Unit capital requirement	=	K/Q
	=	\$2,000,000/150,000,000 bricks per annum
	=	\$0.013333 per brick per annum
	=	\$13.333 per 1,000 bricks per annum

This says that \$13.333 of capital is (required to be) installed per (each) thousand bricks (to be) produced per annum; in other words, in order to have the *capacity* to produce a thousand bricks per annum, a stock of \$13.333 of fixed capital must be installed.

The 'price' of using a dollar of fixed capital in production has to be calculated as the annual rate of depreciation on a straight-line basis (where throughout we assume zero scrap value). That is, where n is the length of life in years of the fixed capital, $1/n$, the proportionate annual rate of depreciation, is the resource price for using a dollar of fixed capital in production; to be consistent with the other notation we shall denote this price as p_K – that is, $p_K \equiv 1/n$. (As explained in Chapter 3 of my book *Profitability, Mechanization and Economies of Scale*, the resource price of capital excludes the financing price for the acquisition of fixed capital because financing is best treated as a matter entirely distinct from production, with the sum of the resource price of fixed capital and the financing price of fixed capital being the rental price of fixed capital.)

Suppose that the length of life of the brick-making plant were five years, giving an annual straight-line proportionate rate of depreciation of 1/5 years = 0.20 per annum = p_K, then the unit capital cost may be calculated as:

Unit capital cost	=	$p_K \times (K/Q)$

$$= 0.20 \text{ per annum} \times (\$2{,}000{,}000/150{,}000{,}000 \text{ bricks per annum})$$

$$= 0.20 \text{ per annum} \times \$0.013333 \text{ per brick per annum}$$

$$= \$0.00267 \text{ per brick}$$

$$= \$2.67 \text{ per } 1{,}000 \text{ bricks}$$

Again, unit capital cost can be more simply understood as annual straight-line depreciation, calculated as $(1/n)K$ – making the substitution of $1/n$ for p_K – or \$400,000 per annum (= (1/5 years) × \$2,000,000) divided by annual output Q of 150,000,000 bricks per annum, so that unit capital cost is straight-line depreciation per unit of output. But it is analytically more useful to understand unit capital cost as the resource price of fixed capital multiplied by unit capital requirement.

Consequently, the unit labour and capital cost combined in this example would be:

$$\text{Unit labour and capital cost combined} = [p_L \times (L/Q)] + [p_K (K/Q)]$$

$$= [\$0.008 \text{ per brick}] + [\$0.00267 \text{ per brick}]$$

$$= \$0.01067 \text{ per brick}$$

$$= \$10.67 \text{ per } 1{,}000 \text{ bricks}$$

One reason for using this simple hypothetical example is that the impact of each of technological change and learning is to reduce *both* unit labour requirement *and also* unit capital requirement simultaneously, and thus (assuming constant p_L and p_K – because in this context we do not wish to be concerned with changes in the prices for inputs) technological change and learning each reduces unit labour and capital cost combined.

Unit total cost is the total resource cost of production divided by the total quantity of output produced. If we denote the annual total cost of production by C and unit total cost by u, then:

$$u \equiv C/Q$$

Whence:

$$C = uQ$$

To illustrate, if the annual total cost for producing 150,000,000 bricks per annum were \$3,000,000 per annum, the unit total cost would be:

u = \$3,000,000 per annum/150,000,000 bricks per annum

= \$0.020 per brick

= \$20 per 1,000 bricks

Unit total cost is the sum of unit materials cost, unit labour cost and unit capital cost. Providing that unit materials cost does not increase (and we shall consider both real world cases where technological change actually *decreases* unit materials cost and also the theoretical impact of learning on *reducing* unit materials cost), then unit total cost will be decreased by the impact of technological change and learning, and this is one of the most important economic outcomes of technological change and learning.

1.3 Profit, the unit net margin and the efficiency rate of profit

Profit is the difference between total receipts from sales and total costs of production. Profit is always measured on an annual basis, and this means that all flows relating to the calculation of profit should likewise be on an annual basis. Annual total receipts from sales is determined by multiplying the price of the product by the annual quantity produced and sold, Q. If we denote the price for a unit of the product, or 'product price', by p_Q then annual total sales receipts is $p_Q Q$. To illustrate, if the price for bricks were \$0.022 per brick then annual total sales receipts would be:

$p_Q Q$ = \$0.022 per brick × 150,000,000 bricks per annum

= \$3,300,000 per annum

Annual profit, denoted by Π, may then be calculated as the difference between annual total sales receipts and annual total cost:

$\Pi = p_Q Q - uQ$

= \$0.022 per brick × 150,000,000 bricks per annum – \$0.020 per brick × 150,000,000 bricks per annum

= \$3,300,000 per annum – \$3,000,000 per annum

$$= \$300{,}000 \text{ per annum}$$

However, there is another important way to calculate annual profit and this may be seen by factorizing the equation for profit as follows:

$$\Pi = p_Q Q - uQ$$

$$= (p_Q - u)Q$$

The expression $(p_Q - u)$ is the money margin of product price over unit total cost and is known as the 'unit net margin'; 'margin' to signify the absolute or money *difference* between product price and unit total cost; 'unit' to signify that this is the margin *per unit of output*; and 'net' because the cost of depreciation is included in unit total cost and profit is calculated net of depreciation so that the unit net margin is the profit per unit of output.

The last line of this equation may be called the 'profit flow equation' because it expresses the annual flow of profit as the product of the unit net margin and the annual quantity of output produced and sold. The profit flow equation is useful because we can see that when technological change or learning reduces unit total cost, u, then, for a constant product price p_Q and a constant annual quantity of output Q, the annual profit will increase because the unit net margin increases.

To illustrate, suppose the price for bricks, p_Q, were \$0.022 per brick with a unit total cost, u, of \$0.020 per brick. Then the unit net margin is (\$0.022 per brick – \$0.020 per brick) = \$0.002 per brick. Annual profit is the product of the unit net margin and the annual quantity of bricks produced and sold:

$$\Pi = (p_Q - u)Q$$

$$= (\$0.022 \text{ per brick} - \$0.020 \text{ per brick}) \times 150{,}000{,}000 \text{ bricks per annum}$$

$$= \$0.002 \text{ per brick} \times 150{,}000{,}000 \text{ bricks per annum}$$

$$= \$300{,}000 \text{ per annum}$$

If unit total cost were to be reduced to \$0.01709 per brick (the case which is to be discussed in the next chapter), the unit net margin would be increased to (\$0.022 per brick – \$0.01709 per brick) = \$0.00491 per brick and annual profit would increase to \$0.00491 per brick × 150,000,000 bricks per annum = \$736,500 per annum. (In passing, note how the proportionately small reduction in unit total cost results in a proportionately very large increase in annual profit. Whenever a variable, such as profit, is determined as the difference between two determining variables, then a relatively small change in either of the determining variables will lead to a relatively large change in the determined variable.)

However, although the annual flow of profit is important, production managers are more likely to be concerned with profitability or the rate of profit; that is, with the flow of profit relative to the capital used. (The various concepts of, and classification scheme explicating, rates of profit are explained in Chapter 2 of my book *Profitability, Mechanization and Economies of Scale*.) Keeping to our measure of the acquisition cost of capital installed, K, a rate of profit per annum can be defined and calculated as Π/K. For reasons to be explained, I call this rate of profit the 'efficiency rate of profit' and we may derive a useful expression for the efficiency rate of profit by substituting the profit flow equation for the annual profit and rearranging, to give the efficiency rate of profit as the unit net margin divided by unit capital requirement; thus:

$$\Pi/K = ((p_Q - u)Q)/K$$

$$= (p_Q - u)/(K/Q)$$

To illustrate for an annual profit of \$300,000, we can either calculate the efficiency rate of profit as annual profit divided by the acquisition cost of capital installed:

$$\Pi/K = \$300{,}000 \text{ per annum}/\$2{,}000{,}000$$

$$= 0.15 \text{ per annum}$$

$$= 15 \text{ per cent per annum}$$

or we can calculate the efficiency rate of profit as the unit net margin divided by the unit capital requirement:

$$(p_Q - u)/(K/Q) = (\$0.022 \text{ per brick} - \$0.020 \text{ per brick})/(\$2{,}000{,}000/150{,}000{,}000 \text{ bricks per annum})$$

$$= (\$0.022 \text{ per brick} - \$0.020 \text{ per brick})/\$0.013333 \text{ per brick per annum}$$

$$= \$0.002 \text{ per brick}/\$0.013333 \text{ per brick per annum}$$

$$= 0.15 \text{ per annum}$$

$$= 15 \text{ per cent per annum}$$

It is analytically more useful to express the efficiency rate of profit in this alternative way as the ratio between the unit net margin and the unit capital requirement, and this is the way we shall mainly view the efficiency rate of profit.

The importance of all this is that each of technological change and learning has two impacts on efficiency: one is to increase the unit net margin by reducing unit total cost (here we assume a constant product price); the other is to reduce unit capital requirement. Consequently, technological change and learning by increasing the numerator and by decreasing the denominator of the efficiency rate of profit are each capable of having a substantial positive impact on profitability. This is why we use the efficiency rate of profit to measure the impact of technological change and learning – profitability and increased profitability being matters of great practical concern to businesses.

2 Neutral Technological Change and Multifactor Productivity

2.1 Technical change, process innovation, and technological change

The term 'technical change' can be simply defined as any change in knowledge about production: about methods of production, about products, or about inputs to making products. Technical change results in both inventions and innovations. An invention is an initial idea and its implementation in a non-commercial setting (such as a laboratory or workshop). An invention is often based on patentable new knowledge.

An innovation is the successful implementation of an invention (patentable new knowledge) in a commercial setting from production to selling. Innovations may concern either the process of producing goods and services – *how* a good or service is produced – or the products (goods and services) themselves – *what* good or service is produced. If the innovation concerns how (the manner in which) a good or service may be produced, we shall call it a 'process innovation'; if it concerns what (type of) good or service may be produced, we shall call it a 'product innovation'.

This book is concerned with efficiency in production, and so deals only with process innovations (even though process innovations are often determined by product innovations). The classification of process innovations is shown in Figure 2.1.

We make a distinction as to whether the process innovation involves, as a central feature, a physical alteration to the fixed capital (plant and equipment) or to the materials (intermediate products) used in production, on the one hand, or whether it does not involve such a physical alteration, on the other hand. If it involves such a physical alteration as a central feature we shall call it 'technological change', and if it does not involve such a physical alteration as a central feature we shall call it a 'change in technique'. (The distinction

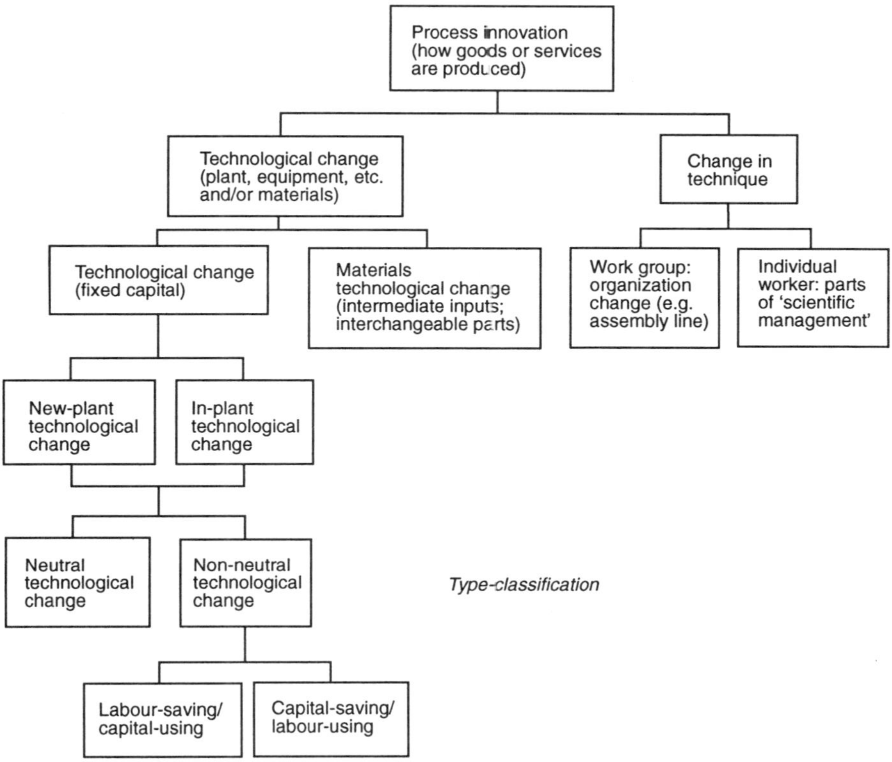

Figure 2.1 The classification of process innovations

between technological change and change in technique is somewhat arbitrary because the phrase 'as a central feature' involves some judgement; however, there are instances at each end of the spectrum which can reasonably be taken to be quite clear, and where the boundary is drawn in between is less important.)

A change in technique may concern either a work group or an individual worker. A change in technique concerning a work group may be called organization change. One classic example of organization change would be the introduction of the assembly line by Henry Ford.

A change in technique concerning an individual worker forms part of what is known as 'scientific management' stemming from the pioneering work of Frederick Winslow Taylor and Frank Gilbreth. Such process innovations often comprise some ergonomic change aimed at improving task performance by increasing the accuracy and speed of operation and/or by reducing fatigue. A classic example of this category of process innovation would be Gilbreth's improvements to bricklaying by using a scaffold with an adjustable platform for holding a packet of properly arranged bricks.

Change in technique is a large but somewhat specialized subject and will not be further discussed. Note that we exclude from change in technique, and therefore also exclude from process innovation, any improvement due to learning: learning involves improved operation of a physically unchanged plant by the same workforce where the improvement arises simply from experience of production and from no other causes. Learning in this sense is an extremely common, widely prevalent, and perpetually on-going condition, so learning will be discussed as an entirely separate topic in the second part of this book.

Technological change is a process innovation involving, as a central feature, a change to fixed capital or to intermediate inputs.

A classic example of technological change involving a change to intermediate inputs would be the introduction of interchangeable parts (usually associated with Eli Whitney of cotton gin fame who began making muskets for the United States government using interchangeable parts; this method of production subsequently became known to contemporaries worldwide as the 'American system' of manufacture).

If the technological change involves intermediate inputs we shall call it 'materials technological change'. This book will not consider materials technological change because of the very many different types of materials which may be used as inputs. However, the importance of materials technological change is immense and this should be clearly acknowledged.

We shall reserve the unqualified term 'technological change' to refer to a *process innovation involving, as its central feature, physical alteration to fixed capital.*

Under the restricted heading of 'technological change' it is useful to make a distinction as to whether an alteration requires different plant or equipment or whether an alteration merely involves modifying or extending existing plant or equipment.

If an alteration involves different plant or equipment, we shall call it 'new-plant technological change'; if an alteration involves modifying or extending existing plant or equipment, we shall call it 'in-plant technological change'.

It may be difficult to draw a precise boundary between new-plant technological change and in-plant technological change because the boundary between what counts as 'different' and what counts as 'modified' is not clear-cut. However, unambiguous examples exist at either extreme.

Both new-plant technological change and in-plant technological change usually have the same basic economic outcome: namely, a reduction in unit input requirements and thence a reduction in unit costs.

Either new-plant technological change or in-plant technological change can be further classified as to the type of change involved. This type-classification concerns the behaviour of the ratio of fixed capital input to labour input at that position where the unit labour and capital cost combined is at a minimum. Because of its reference to fixed capital, this type-classification is confined to technological change involving fixed capital.

There are two such types of technological change: 'neutral' technological change; and 'non-neutral' technological change.

The basis of the type-classification – the criterion used to distinguish between the two – is whether or not the technological change leaves unaltered the ratio of fixed capital to labour *at a constant, unchanged, factor price ratio*. (Throughout, it must be understood that any reference to the capital–labour ratio is a reference to that ratio at which the unit cost of labour and capital combined is at a minimum. The term 'capital-intensity' may also be used to refer to this capital–labour ratio.) A technological change which does *not* alter the capital–labour ratio when the factor price ratio is constant is type-classified as being neutral technological change. A technological change which *does* alter the capital–labour ratio when the factor price ratio is constant is type-classified as being non-neutral technological change.

This is the definition of 'neutral'/'non-neutral' given by W.E.G. Salter, *Productivity and Technical Change* (Cambridge University Press, 1960), pp. 31–4. A converse and almost equivalent definition was given by J.R. Hicks in *The Theory of Wages* (1932), pp. 121–2; namely that the ratio of factor marginal value products is not altered at the same capital–labour ratio. The converse definition is usually referred to as 'Hicks neutral technological change', but, as Salter convincingly argues, Hicks' definition is less useful for the purpose of analysing productivity change at the plant level, and so we shall follow Salter's definition and we shall not use Hicks' definition.

As the next chapter explains, non-neutral technological change may itself be divided among two sub-types: labour-saving/capital-using non-neutral technological change; and capital-saving/labour-using non-neutral technological change. The former is the one which most frequently occurs; the latter is uncommon.

The distinction between neutral and non-neutral technological change is a matter of great theoretical and practical importance. Salter's argument was that, at the plant or industry level, we must be concerned with the impact of technological change (at the same factor price ratio) on the capital-intensity of production and thence on different rates of productivity change among the different factor inputs. The reason why non-neutral technological change is a matter of practical importance is that it leads to an enhanced rate of labour productivity growth; that is, under labour-saving/capital-using non-neutral technological change, labour productivity tends to be greater, and labour productivity growth tends to be faster, than would be the case under neutral technological change. The theoretical importance of this does not require comment. The practical importance arises from the fact that most technological change is, and has historically been, non-neutral.

2.2 Neutrality and the production function

The fundamental economic feature of technological change (process innovation) is that a greater quantity of output may be produced with a lesser quantity of inputs. In other words, technological change simultaneously reduces both unit labour requirement and also unit capital requirement. Another feature of technological change is that it may have a different impact on each of the unit input requirements. In order to analyse these features of technological change we need to use a production function which relates inputs to output. The production function which will be used throughout this book is the Cobb–Douglas production function with constant returns to scale. The justification for using this relatively simple form of production function is that these fundamental features of technological change can be described and analysed in the simplest possible way without complex mathematics obscuring the economics; the justification for using a constant returns to scale production function is that the analysis of technological change needs to be kept separate from the effects of increasing returns to scale. (Increasing returns to scale is analysed in my book *Profitability, Mechanization and Economies of Scale,* as are mechanization and the cost-minimizing conditions of production.)

The Cobb–Douglas production function relates the quantity of output per period (generally per annum, and we shall throughout keep to a production period of one year), measured generally in *physical* units (such as tons or ton-miles) – but constant-price money measures of output may also be used – and denoted by Q, to quantities of factor inputs. (Strictly speaking, this should be called the *factor input frontier production function*: 'factor' because the function deals only with the factor inputs of labour and fixed capital, and assumes all other inputs are adjusted as required; and 'frontier' because the function theoretically relates to only those combinations of inputs which are technically efficient, meaning that it is not possible to reduce the quantity of any one input without increasing the quantity of another input while still continuing to produce the same quantity of output per period.)

The factor inputs are also measured in units of, or relating to, a physical quantity of the input. For the labour input, a generally used measure is the quantity of labour-hours used per annum. For the input of fixed capital, the 'quantity' has necessarily to be measured by the money acquisition cost, *at constant prices*, of the fixed assets comprising the plant. (The reason and rationale for this approach to capital input, and for the resource 'price' of capital to be taken as the proportionate annual rate of (straight-line) depreciation, are explained in *Profitability, Mechanization and Economies of Scale*.) The quantity of input of labour-hours per annum will be denoted in this book by L, and the acquisition cost at constant prices of the fixed capital stock will be denoted by K.

The production function requires what I call an 'adjustment factor', denoted by Z. In the first place, the adjustment factor permits the production function relating inputs to output to 'work' as a matter of arithmetic. In the second place, in the analysis of technological change, Z, if permitted to change (increase), enables us to analyse the fundamental economic feature of technological change, namely an increased quantity of output in relation to the same or reduced quantities of factor inputs.

The production function has two parameters, a partial elasticity of the quantity of output with respect to the quantity of labour input (when the quantity of capital is held constant) – this factor input partial elasticity to be denoted by a; and a partial elasticity of output with respect to the quantity of fixed capital installed (when the quantity of labour input is held constant) – this factor input partial elasticity to be denoted by b. That is, a measures the *proportionate* response of Q to a *proportionate* change in L (when K is constant), and b measures the *proportionate* response of Q to a *proportionate* change in K (when L is constant).

The restriction of the analysis (in this book) to technological change means that constant returns to scale is assumed throughout, and this requires that the sum of the factor input partial elasticities be equal to 1. Accordingly $a + b = 1$ (and note that therefore $-(a/b) - 1 = -1/b$).

The factor input frontier production function can thus be written:

$$Q = Z L^a K^b$$

The implication of the Salter definition of neutral technological change can now be explained using the production function as follows.

The isoquant expresses the total quantity of capital input, K, as a function of the total quantity of labour input, L, for any given value of the total quantity of output per annum, Q. The isoquant for this production function may be derived by manipulating the production function to give K as a function of L:

$$K = (Q/Z)^{1/b} L^{-a/b}$$

Along the isoquant, the marginal physical rate of substitution between capital and labour, $\mathrm{d}K/\mathrm{d}L$, is:

$$\mathrm{d}K/\mathrm{d}L = -(a/b)\,(Q/Z)^{1/b} L^{-(a/b)-1}$$

$$= -(a/b)\,(Q/Z)^{1/b} L^{-1/b}$$

The condition for cost minimization (with respect to factor inputs) is that the marginal money rate of substitution be equal to (–)1, and this in turn requires that the marginal physical rate of substitution be equal to the negative of factor price ratio, where the factor price ratio is defined as the price of a unit of labour (where the unit of labour input is a labour-hour, and therefore the price of a unit of labour is measured as the 'wage rate' per labour-hour), denoted by p_L, divided by the resource price of a unit of capital (which 'price' is the proportionate annual rate of depreciation), denoted by p_K. That is, the condition for cost minimization is either:

$$(p_K\,\mathrm{d}K)/(p_L\,\mathrm{d}L) = (-)1$$

which is the condition that the marginal money rate of substitution be equal to (–)1, or (equivalently):

$$\mathrm{d}K/\mathrm{d}L = -(p_L/p_K)$$

which is the condition that the marginal physical rate of substitution be equal to the (negative of the) factor price ratio. By substitution for $\mathrm{d}K/\mathrm{d}L$ this latter condition is:

$$-(a/b)\,(Q/Z)^{1/b} L^{-1/b} = -(p_L/p_K)$$

This cost-minimizing condition can be solved for that value of the quantity of labour input which gives the minimum cost quantity of labour input for any value of Q, this minimum cost quantity of labour input to be denoted by L^*:

$$L^* = ((a/b)/(p_L/p_K))^b(Q/Z)$$

By substitution into the isoquant of the right-hand side of this expression for L^* and simplifying, the corresponding value of the cost-minimizing quantity of capital input, K^*, can be derived as:

$$K^* = ((a/b)/(p_L/p_K))^{-a}(Q/Z)$$

From this the capital–labour ratio for minimum cost production, to be denoted by $(K/L)^*$, can be obtained by dividing the expression for L^* into the expression for K^* and simplifying:

$$(K/L)^* = (p_L/p_K)/(a/b)$$

If the factor price ratio, p_L/p_K, is constant, then $(K/L)^*$ remains unaltered only if the value of the ratio of the exponents to the production function, a/b, remains unchanged.

Therefore, the condition for neutral technological change is that the value of the ratio a/b remains the same. In these conditions the capital-intensity of production does not vary with technological change, and this is, accordingly, 'neutral' technological change; that is, neutral with regard to the capital–labour ratio (at the minimum cost position).

2.3 Neutral technological change

Technological change is a complex phenomenon with many different facets. Usually, but not invariably, a process innovation has as its primary objective the reduction of unit cost through the reduction of unit input requirements. That is, neutral technological change increases the *quantity* of output in relation to inputs, or reduces the *quantity* of inputs in relation to output, or both. These are the features of technological change which may be analysed through a production function. However, process innovations may be directed towards improving the *quality* of a product or the *safety* of a process or towards other goals (such as protection of the environment), and we should recognize that these other facets exist. It is difficult to analyse these other features of technological change through a production function (in the case of quality improvement a quality adjusted measure of output could be used), but understanding the production function analysis of the main features of technological change puts us in a better position to understand the impact of these other features. Accordingly, the main focus of our attention will be on how technological change affects unit input requirements and particularly unit factor input requirements.

To analyse this primary objective of technological change, we shall use the factor input frontier production function: $Q = Z L^a K^b$.

Throughout our explanation of technological change (whether neutral or non-neutral) we use a production function with constant returns to scale, and this requires that the sum of the exponents be equal to 1: $a + b = 1$. It is possible to have a production function where this restriction does not hold but we will not be concerned with this matter in this book.

Because technological change is something that happens through time, it is necessary to introduce a time subscript, t, into the production function, and to subscript all variables with regard to time. As we shall see, this means that we must also subscript the adjustment factor Z with regard to time. This means that Z becomes a variable in the production function instead of being a (constant) parameter of the production function. But this makes no essential difference to the functioning of Z as an adjustment factor and, as we shall see, it does permit us to capture the essential economic feature of technological change.

In the context of measuring and analysing technological change, the time subscript on the variables requires a precise and peculiar interpretation because, for technological change, the contrast should be thought of as a contrast between plants of differing 'vintages' rather than between the 'same' plant at two different time periods. That is, in this section Q_t should be seen as denoting the physical quantity of (annual) output from a (technically efficient) plant of vintage t, which means a plant using the most advanced technology available at t. Consequently and likewise, Q_{t+1} should be seen as denoting the physical quantity of (annual) output from a (technically efficient) plant of vintage $t + 1$ using the even better technology than that available at t.

Subsequently, this 'vintage' interpretation of t may be relaxed, but it remains the interpretation basic to technological change. That is, the analysis of technological change should basically be in terms of the comparison of two entirely distinct technically efficient plants: one plant embodying or using the old, superseded technology; the other plant built from scratch on a 'green-field' site and embodying or using the new technology.

Thus, for the period t (with the subscript for flows and stocks being likewise appropriately interpreted), we have as the production function for a plant using technology of vintage t:

$$Q_t = Z_t L_t^a K_t^b$$

We have as the production function for a plant using technology of vintage $t + 1$:

$$Q_{t+1} = Z_{t+1} L_{t+1}{}^{a} K_{t+1}{}^{b}$$

Suppose that between t and $t + 1$ there has been neutral technological change. Suppose that this neutral technological change has permitted a reduction in the quantities of both labour input and capital input, and at the same time has led to a slight increase in the quantity of annual output.

If there had been no technological change, then the production function analysis would lead us to expect a decrease in output corresponding to the decrease in factor inputs. We could denote this corresponding hypothetical output by Q'' and could calculate its value as follows:

$$Q'' = Z_t L_{t+1}{}^{a} K_{t+1}{}^{b}$$

We can argue that the change from Q_t to Q'' would be due to the change in L and K (if the technology did not change), and that therefore the change from Q'' to Q_{t+1} can be ascribed to neutral technological change.

We may thus measure the proportionate impact, or rate, of neutral technological change as the proportionate difference between Q_{t+1} and Q'', as follows:

$$\begin{aligned}(Q_{t+1} - Q'')/Q'' &= (Q_{t+1}/Q'') - 1 \\ &= ((Z_{t+1} L_{t+1}{}^{a} K_{t+1}{}^{b})/(Z_t L_{t+1}{}^{a} K_{t+1}{}^{b})) - 1 \\ &= (Z_{t+1}/Z_t) - 1\end{aligned}$$

Hence the proportionate change in the adjustment factor, Z, measures the proportionate impact on output of neutral technological change. We have no direct observations on Z, but we may compute a value of Z, for any period or vintage t, as follows:

$$Z_t = Q_t/(L_t^{a} K_t^{b})$$

In this calculation of Z, the numerator of the ratio on the right-hand side is simply the quantity of annual output, Q_t.

The denominator in the ratio is the variable $(L_t^{a} K_t^{b})$. This variable can be thought of as, and in fact is, a measure of the *combined factor input*, measured in the form of an index.

The ratio is therefore a ratio between output and combined factor input, so we can identify the ratio as *multifactor productivity*. The concept and measurement of the combined factor input and of its associated multifactor productivity will be more fully explained and illustrated in the sections following.

If we have observations on the variables Q, L, and K, in technically efficient plants, and if we have estimates of the parameters a and b, then it is possible to calculate the value of Z for any period (or vintage) and to take the proportionate change in Z as a measure of the rate of neutral technological change, or rather as a measure of the proportionate impact of neutral technological change on output in technically efficient plants.

Let us see what all this means in terms of an hypothetical arithmetic example and the unit isoquant.

Consider an hypothetical production function for a product, say bricks. Suppose that the production function for bricks is: $Q = Z L^{0.75} K^{0.25}$. Suppose that a brick-making plant produces output Q of 150,000,000 bricks per annum with a labour input, L, of 500,000 labour-hours per annum and with an acquisition cost of capital installed, K, of \$2,000,000.

Assume a factor price ratio p_L/p_K = \$2.40 per labour-hour/0.20 per annum = 12. Consequently the capital–labour ratio at the minimum unit cost position, $(K/L)^*$, is $(p_L/p_K)/(a/b)$ = (\$2.40 per labour-hour/0.20 per annum)/(0.75/0.25) = 12/3 = \$4 per labour-hour per annum. Consequently the hypothetical plant is operating with the minimum cost capital–labour ratio: \$2,000,000/500,000 labour-hours per annum = \$4 of capital installed per labour-hour per annum.

The value of Z for such a minimum cost plant may be calculated from the formula derived from the production function as $Z_t = Q_t/(L_t^{0.75} K_t^{0.25})$:

$$Z_t = 150{,}000{,}000/(500{,}000^{0.75} \times 2{,}000{,}000^{0.25})$$

$$= 212.1320344$$

Suppose that there is a technological change which permits there to be, in a technically efficient plant of vintage $t + 1$ which embodies the new technology, a 20 per cent reduction in each of the factor inputs to L_{t+1} = 400,000 labour-hours per annum and K_{t+1} = \$1,600,000, but with a 10 per cent increase in the quantity of annual output to Q_{t+1} = 165,000,000 bricks per annum.

This hypothetical example deliberately uses a mixture of a decrease in the quantities of the factor inputs and a slight increase in the quantity of annual output because this is a common outcome of technological change. Hence we can refer to 'producing more output with less input' as the primary outcome of technological change and as its fundamental economic feature.

Note that $(K/L)^*$ still remains at \$4 per labour-hour per annum.

Thus for a plant of vintage $t + 1$:

$$Z_{t+1} = 165{,}000{,}000/(400{,}000^{0.75} \times 1{,}600{,}000^{0.25})$$
$$= 291.6815473$$

The impact of the neutral technological change on output can thus be measured as:

$$(Z_{t+1}/Z_t) - 1 = (291.6815473/212.1320344) - 1$$
$$= 1.375 - 1$$
$$= 0.375$$
$$= 37.5 \text{ per cent}$$

Effectively, this says that neutral technological change would permit a 37.5 per cent increase in output for the same quantity of factor inputs.

We shall refer to this 37.5 per cent increase (or any similarly computed proportionate increase) as the 'rate of technological change', but you should remember that this is an abbreviated way of referring to the impact of technological change on the quantity of output capable of being produced by the same quantities of factor inputs.

We need to see what such neutral technological change does to the frontier unit isoquant because this is one important way of representing the impact of technological change.

2.4 The unit isoquant under neutral technological change

The isoquant expresses the total quantity of capital input, K, as a function of the total labour input, L, for any given value of total output per annum, Q; the unit isoquant expresses unit capital requirement, (K/Q), as a function of unit labour requirement, (L/Q). The general formula for the unit isoquant may be derived from the isoquant by dividing the isoquant by Q and multiplying also by $Q^{a/b} \times Q^{-a/b} = 1$ and rearranging to give the requisite unit labour requirement on the right-hand side:

$$(K/Q) = Z^{-1/b} Q^{(1-a-b)/b} (L/Q)^{-a/b}$$

In the case of a production function with constant returns to scale, $a + b = 1$, so that $(1 - a - b) = 0$; and the term in Q alone disappears from the unit isoquant, which is in this case not dependent upon the scale of production.

When there is technological change, Z becomes a variable which changes with the vintage of the technology, so Z must be subscripted with regard to t, and the unit isoquant in the case of constant returns to scale becomes:

$$(K/Q) \quad = \quad Z_t^{-1/b} (L/Q)^{-a/b}$$

Because of its negative exponent, if Z increases, the unit capital requirement, (K/Q), will be smaller for any given value of the unit labour requirement, (L/Q).

Table 2.1 presents the unit isoquant before the neutral technological change and the unit isoquant after the neutral technological change. Table 2.1 also gives the unit cost of labour and capital combined along each unit isoquant, and the minimum unit cost positions are asterisked.

Figure 2.2 graphs (part of) each unit isoquant. We can see that the unit isoquant has shifted downwards and to the left because of the neutral technological change. The downward shift of the unit isoquant is important because it depicts the fundamental economic feature of technological change: namely, being able to produce more output with less input.

It is also useful to represent the impact of neutral technological change through the unit isoquant because we can see what is happening to unit cost at the minimum cost combination of labour and capital inputs. The minimum cost combinations of factor inputs occur at the intersection of the unit isoquants and the capital–labour ratio of $4 of capital installed per labour-hour used per annum; that is, at A and B.

It is most important to note that we must make any comparison *in terms of the minimum cost positions*. This is the only way of properly analysing what is going on. If we compare a minimum cost position with a position that is not minimum cost, then we have a mixture of influences in the comparison and we can sort out this mixture only by reference to minimum cost positions. We can give a theoretically satisfactory description and analysis of what is going on only by comparing minimum cost positions. These important propositions will be illustrated and exemplified in the next section's discussion of the labour productivity form of the production function, to be shown in Figure 2.3.

In Table 2.1, the minimum unit cost for labour and capital combined is $10.67 per thousand bricks before the neutral technological change and decreases to $7.76 per thousand bricks after the neutral technological change. This is a decrease of 27.3 per cent. The decrease in the minimum unit cost for labour and capital combined for this 37.5 per cent rate of technological change

Table 2.1
The shift in the unit isoquant due to neutral technological change (hypothetical data)

Unit isoquant[a]				Unit labour and capital cost combined,[b] \$ per 1,000 bricks, $(p_L(L/Q) + p_K(K/Q))$ 1,000	
Before neutral technological change		After neutral technological change			
Unit labour requirement, labour-hours per brick, L/Q	Unit capital requirement, value of capital installed per brick per annum, \$, K/Q	Unit labour requirement, labour-hours per brick, L/Q	Unit capital requirement, value of capital installed per brick per annum, \$, K/Q	Before neutral technological change	After neutral technological change
0.002000	0.061728	0.002000	0.017269	17.15	8.25
0.002333	0.038873	0.002333	0.010875	13.37	7.77
0.002424	0.034661	0.002424	0.009697	12.75	7.76*
0.002667	0.026042	0.002667	0.007285	11.61	7.86
0.003000	0.018290	0.003000	0.005117	10.86	8.22
0.003333	0.013333	0.003333	0.003730	10.67*	8.75
0.003667	0.010018	0.003667	0.002803	10.80	9.36
0.004000	0.007716	0.004000	0.002159	11.14	10.03
0.004333	0.006069	0.004333	0.001698	11.61	10.74
0.004667	0.004859	0.004667	0.001359	12.17	11.47

(*a*) Values of (K/Q) calculated from the formula for the unit isoquant under constant returns to scale:

$$(K/Q) = Z^{-1/b} (L/Q)^{-a/b}$$

with $Z = 212.1320344$ (before neutral technological change) and $Z = 291.6815473$ (after neutral technological change); with $a = 0.75$ and $b = 0.25$; and using the following unrounded values of $L/150{,}000{,}000$:
L = 300,000; 350,000; 400,000; 450,000; 500,000; 550,000; 600,000; 650,000; 700,000;
with the interpolation in the third row of 400,000/165,000,000.

(*b*) For p_L = \$2.40 per labour-hour, and p_K = 0.20 per annum. Asterisk, *, denotes minimum cost position.

Source: see text.

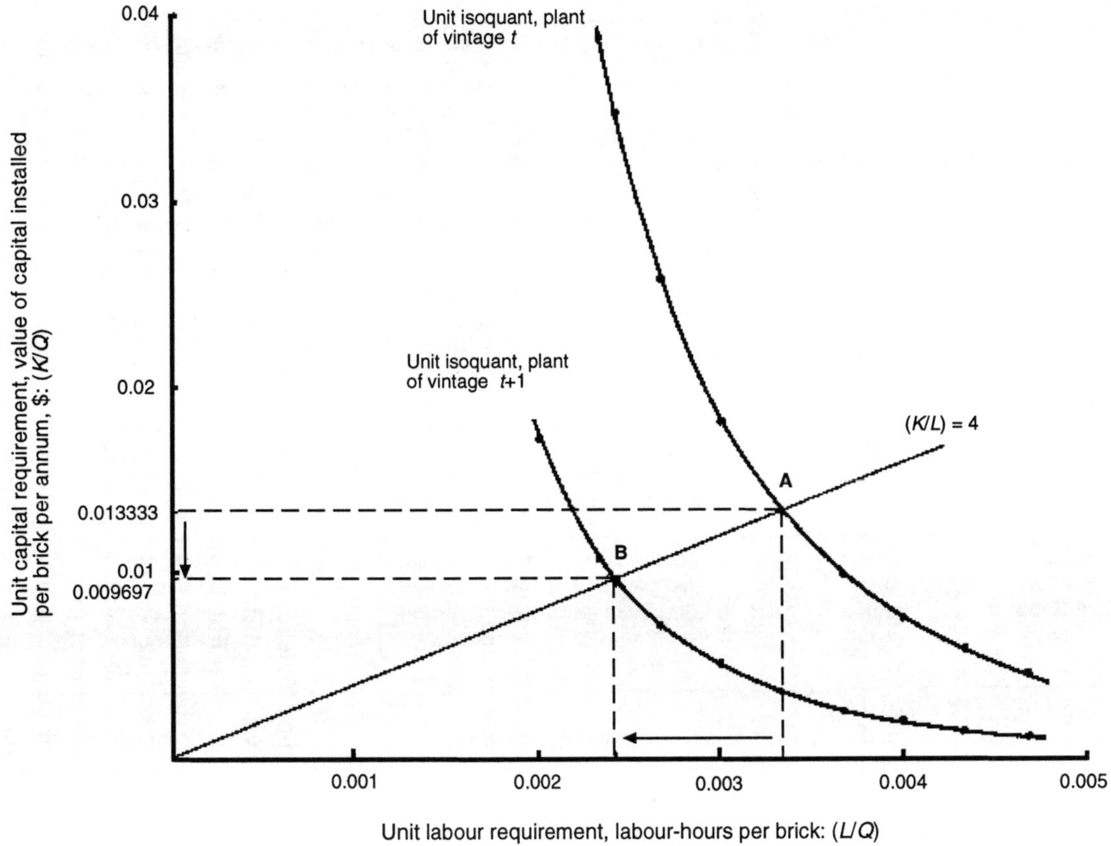

Figure 2.2 The unit isoquant under neutral technological change

can also be computed as (1/1.375) – 1. In other words, the decrease in the unit cost of labour and capital combined is the inverse of the rate of neutral technological change.

If we were to assume a constant unit materials cost of $9.33 per thousand bricks, the unit total cost for bricks, at the minimum cost position, would be $20 per thousand bricks before neutral technological change (= $10.67 per thousand bricks + $9.33 per thousand bricks) and $17.09 per thousand bricks after neutral technological change (= $7.76 per thousand bricks + $9.33 per thousand bricks).

Assuming a constant product price for bricks of $22 per thousand bricks, this change in unit total cost would mean an increase in the unit net margin from $2.00 per thousand bricks before neutral technological change (= $22 per thousand bricks – $20 per thousand bricks) to $4.91 per thousand bricks after neutral technological change (= $22 per thousand bricks – $17.09 per thousand bricks), so the unit net margin would be increased two-and-a-half-fold because of the technological change!

But this is not all that has to be considered. Unit capital requirement (at the minimum cost position) is decreased from $13.333 per thousand bricks per annum before neutral technological change to $9.697 per thousand bricks per annum after neutral technological change (Table 2.1).

As explained in Chapter 1, the efficiency rate of profit is the ratio between the unit net margin and the unit capital requirement, so (at the minimum cost position) the efficiency rate of profit increases from 15 per cent per annum before neutral technological change to 50.6 per cent per annum after neutral technological change. The efficiency rate of profit rises due to *both* the increase in the unit net margin and *also* the decrease in unit capital requirement. In this hypothetical example, neutral technological change results in a more than trebling of the efficiency rate of profit!

This serves to indicate the economic power of technological change. Of course, in the real world we would have to query the assumption of a constant product price, because it is more than likely that the price charged would be reduced, thus sharing with product purchasers some at least of the gains from technological change. But such speculation about the subsequent market behaviour of product price is secondary to the analysis of the impact of technological change at the plant level on production costs and unit input requirements.

2.5 Multifactor productivity: theory

If we give the appropriate interpretation to the time subscripts, so that the contrast between Z_t and Z_{t+1} is a contrast between plants of different 'vintages', each using a different technology to produce the same type of output, and each plant operating at full capacity and at the same scale of output (so that there is no possibility of influences from economies of scale) and producing at minimum unit total cost, then the proportionate difference in Z between the economically efficient plants can be interpreted as measuring the impact of technological change.

However, if these conditions are not met, then differences, or changes, in Z may have to be ascribed to other influences (such as an increase in capacity utilization or economies of scale or a movement from technically inefficient production towards technically efficient production) as well as to technological change. In the real world, change in multifactor productivity is caused by more things than simply technological change, although in the long run technological change is likely to be the predominant cause of an increase in multifactor productivity.

Measured changes in Z (whatever their cause) are of interest because such changes indicate the extent to which an economy, or part thereof, is able to produce more with less factor input. In this context, the adjustment factor Z has been called 'multifactor productivity', and increases in Z are referred to as 'multifactor productivity growth'. In this section we explain the concept of multifactor productivity and in the next section, using data on manufacturing industry in the United States for the period 1948 to 1981, we show how changes in multifactor productivity may be calculated.

We are generally accustomed to dealing with single input productivity measures, such as labour productivity which is defined as quantity of output per unit of labour input, (Q/L). However, suppose that after you contemplate with approval an increase in labour productivity in a plant, someone criticizes the plant with the following objection: 'That increase in labour productivity does not tell you very much; that plant has increased its labour productivity simply by installing excessive, unnecessarily large, amounts of fixed capital.'

The important implication of this criticism is that the high productivity of one input is the result of the low productivity, or uneconomic use, of another input. But this criticism is more complex than might at first sight appear, as we shall subsequently show using two possible scenarios for labour productivity change.

In any such case it becomes necessary to consider the productivity of more than one input, so that we can see whether the high productivity of one input is being counteracted by (a causally related) low productivity of another input. Consequently, rather than look at the productivity of each input separately, it is necessary to consider the productivity of combined inputs, especially the combined inputs of labour and capital.

The measure of combined labour and capital productivity is generally called *multifactor productivity*. (Sometimes this may be referred to as 'total factor productivity'.)

Multifactor productivity may be measured in either of two ways.

In the first place, we may, as already explained, measure Z_t as $Q_t/(L_t^a K_t^b)$. The adjustment factor, Z, is a ratio between output in the numerator and a combined measure of factor inputs, labour and capital, in the denominator. The combined factor input is measured by $L_t^a K_t^b$. This latter expression is effectively an index of the combined quantities of L and K. When this index of combined factor input is divided into a measure of output then the ratio provides a measurement in index values of multifactor productivity. Measuring Z in this way requires observations on the variables Q, L, and K, together with estimates of the parameters a and b.

In the second place, we may use the labour productivity form of the constant returns to scale production function to calculate a value of Z.

The labour productivity form of the production function gives labour productivity, (Q/L), as a function of the capital–labour ratio, (K/L). The labour productivity form of the production function may be derived from the constant returns to scale production function where $a + b = 1$, so $a - 1 = -b$. The derivation is as follows:

$$
\begin{aligned}
(Q/L)_t &= Z_t L_t^{a} K_t^{b} L_t^{-1} \\
&= Z_t L_t^{a-1} K_t^{b} \\
&= Z_t L_t^{-b} K_t^{b} \\
&= Z_t (K/L)_t^{\,b}
\end{aligned}
$$

The last line of this equation expresses the labour productivity form of the production function and from this we may derive:

$$Z_t = (Q/L)_t/(K/L)_t^b$$

Measuring Z in this way requires observations on labour productivity, (Q/L), and the capital–labour ratio, (K/L), together with an estimate of the parameter b.

Now consider carefully the following quotation from a publication of the United States Department of Labor Bureau of Labor Statistics, one of the first publications to present official estimates of multifactor productivity:

> This bulletin presents a recently constructed measure of productivity – multifactor productivity ... The index of multifactor productivity measures annual change in output per unit of combined labor and capital input. This is mathematically equivalent to subtracting the effects of annual rates of change in capital per [labour-] hour from the annual rates of change in output per [labour-] hour of all persons [employed]. Thus, the multifactor productivity measure differs from the familiar BLS [Bureau of Labor Statistics] measure of output per [labour-] hour of all persons [employed] in that it *excludes the effects of capital intensity*. ... the multifactor productivity index can be interpreted as one of a number of indicators of the economic progress of the U.S. economy because it shows the growth in output that has been obtained from a given amount of [combined] resources (capital and hours of labor), or, conversely, the reduction over time in the quantity of these resources used to produce a unit of output. ... the aggregate measure of output per [labour-] hour of all persons [employed] reflects many influences, such as the amount of capital per unit of labor, ... capacity utilization, and the organization of production. ... The BLS index of multifactor productivity, which measures output per unit of combined labor and capital, is, in fact, an index of output per [labour-] hour of all persons [employed] *adjusted for the influence of capital per [labour-] hour*. (United States Department of Labor Bureau of Labor Statistics, *Trends in Multifactor Productivity, 1948–81*, Bulletin 2178, September 1983 (hereafter referred to as Bulletin 2178), pp. 1 and 16, italics added; see also J.A. Mark and W.H. Waldorf, 'Multifactor productivity: a new BLS measure', *Monthly Labor Review*, December 1983.)

An important point is that the measure of multifactor productivity *excludes*, or is adjusted for, the impact on labour productivity of changes in the capital–labour ratio. This can be explained in terms of the labour productivity form of the production function depicted in Figure 2.3. This explanation continues the example discussed in the previous section and illustrated by the downward movement of the unit isoquants shown in Figure 2.2. Suppose we have two labour productivity production functions where $b = 0.25$. Before neutral technological change, the labour productivity production function is:

$$(Q/L) \quad = \quad 212.1320344\ (K/L)^{0.25}$$

After neutral technological change, the labour productivity production function is:

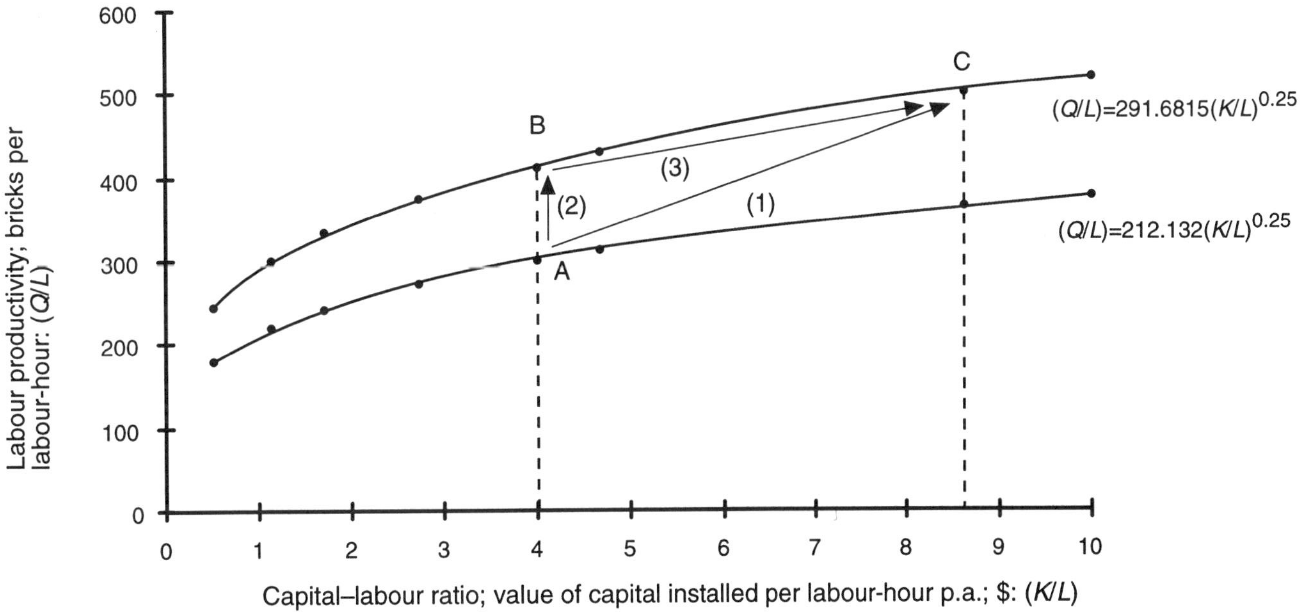

Figure 2.3 The labour productivity production function under neutral technological change

$$(Q/L) = 291.6815473\,(K/L)^{0.25}$$

Each of these production functions is drawn in Figure 2.3. The effect of the neutral technological change, measured by the 37.5 per cent increase in the adjustment factor Z from 212.1320344 to 291.6815473, is to shift the labour productivity form of the production function upward. This is the counterpart to the downward shift in the unit isoquant. Suppose we then have the following observations for t and $t + 1$:

	t	$t + 1$
Labour productivity, bricks per labour-hour, (Q/L)	300	500.00
Capital–labour ratio, \$ of capital installed per labour-hour per annum, (K/L)	4.00	8.63

For the following explanation, we need to consider two scenarios. First, we shall consider the scenario where the factor price ratio is throughout constant at \$2.40 per labour-hour/0.20 per annum = 12. With the ratio of the exponents to the production function remaining the same at 0.75/0.25 = 3, the capital–labour ratio at the minimum cost position, $(K/L)^* = (p_L/p_K)/(a/b)$, stays at \$4 of capital installed per labour-hour used per annum. This means that the position at $t + 1$ with its higher capital–labour ratio would *not* be a minimum cost position, and the criticism about 'excessive, unnecessarily large, amounts of fixed capital' would be justified.

Second, we shall consider the scenario where, after the technological change, the factor price ratio changes to \$5.18 per labour-hour/0.20 per annum = 25.9, and so the capital–labour ratio at the minimum cost position becomes \$8.63 of capital installed per labour-hour used per annum (because $(K/L)^* = (\$5.18/0.20)/(0.75/0.25) = 8.63$). The change in the factor price ratio means that the position at $t + 1$ with its (now appropriately) higher capital–labour ratio *would* be a minimum cost position, and the criticism about 'excessive, unnecessarily large, amounts of fixed capital' would not be justified.

The pair of observations for t is represented by point A in Figure 2.3; the pair of observations for $t + 1$ is represented by point C in Figure 2.3. We can see that the increase in labour productivity is due partly to the upward shift of the labour productivity production function (the increase in Z) and partly to the movement along the labour productivity production function (because of the increase in (K/L)).

Because we are now dealing with the broader matter of multifactor productivity change, rather than with the more restricted matter of technological change (although the former is largely, in the long run, determined by the latter), we shall in what follows refer to multifactor productivity change.

Using the labour productivity form of the production function, we can apportion the increase in labour productivity into two portions as follows.

Compute the (hypothetical) value of labour productivity as it would have been had there been a change in multifactor productivity but had there been no change in the capital–labour ratio of \$4 per labour-hour per annum. Denote this (hypothetical) value of labour productivity as $(Q/L)'$. Then:

$$(Q/L)' = 291.6815473 \times 4^{0.25}$$

$$= 412.50 \text{ bricks per labour-hour}$$

$(Q/L)'$ is at B in Figure 2.3. Using the annotations in Figure 2.3, we have the following proportionate changes (each proportionate change being given in the form of a ratio which is simply another way of presenting a proportionate change):

(1) Total increase in labour productivity (ratio form) — From A to C

$$\frac{500 \text{ bricks per labour-hour}}{300 \text{ bricks per labour-hour}} = 1.667$$

(2) Increase in labour productivity due to multifactor productivity increase (ratio form) — From A to B

$$\frac{412.50 \text{ bricks per labour-hour}}{300 \text{ bricks per labour-hour}} = 1.375$$

(3) Increase in labour productivity due to increase in capital–labour ratio (ratio form) — From B to C

$$\frac{500 \text{ bricks per labour-hour}}{412.50 \text{ bricks per labour-hour}} = 1.212$$

Consequently we may write:

$$\text{Increase in labour productivity due to multifactor productivity increase (ratio form)} = \frac{\text{Total increase in labour productivity}}{\text{Increase in labour productivity due to increase in capital–labour ratio}}$$

$$= 1.667/1.212$$

$$= 1.375$$

Hence the measure of the increase in multifactor productivity 'excludes the effects of capital intensity' and is an index of labour productivity 'adjusted for the influence of capital per [labour-] hour'. (The use of the phrase 'subtracting the effects of ...' in the quotation from Bulletin 2178 is not technically strictly correct, as the preceding calculations show, but the non-technical phrase serves to convey the basic idea. Strictly and technically the 'subtracting', or the 'adjusting', must be done by dividing using the ratio form of relative change.)

The importance of this result is that at the level of an individual plant we would be concerned with economic efficiency and this means that the unit cost of production should be at a minimum. Labour productivity can be higher in a plant if the plant has an increased capital-intensity of production. In other words, it is possible to 'buy' increased labour productivity by increasing the amount of capital installed per labour-hour used per annum. But the real question is whether having such higher labour productivity is 'worthwhile'.

In our first scenario (where the minimum cost capital–labour ratio remains at \$4), B is a minimum cost position, and C is not a minimum cost position. The arithmetic of this can be seen in Table 2.2, which gives the unit costs for positions A, B and C under each of the two scenarios.

Under the first scenario, after the technological change, position B gives the minimum unit labour and capital cost combined of \$7.76 per thousand bricks. If a plant (after technological change) were to have a capital–labour ratio higher than that at B, say at C, then although there would be a reduction in unit labour cost consequent upon the higher labour productivity, the increase in unit capital cost would be greater and so unit labour and capital cost combined would increase. In the first scenario's C–B comparison, unit labour cost would be less by \$1.02 per thousand bricks, but unit capital cost would be greater by \$1.51 per thousand bricks, thus making for a difference in unit labour and capital cost combined of +\$0.49 per thousand bricks.

Table 2.2
Unit costs under different factor price ratios before (A) and after (B and C) technological change (hypothetical data)

	Scenario 1: p_L = \$2.40 per labour-hour; p_K = 0.20 per annum; $(K/L)^*$ = \$4 of capital installed per labour-hour p.a.			Scenario 2: p_L = \$5.18 per labour-hour; p_K = 0.20 per annum; $(K/L)^*$ = \$8.63 of capital installed per labour-hour p.a.		
	Unit cost, \$ per 1,000 bricks[(a)]			Unit cost, \$ per 1,000 bricks[(a)]		
Position in Figure 2.3	Labour cost	Capital cost	Labour and capital cost combined	Labour cost	Capital cost	Labour and capital cost combined
A	8.00	2.67	10.67	17.26	2.67	19.93
B	5.82	1.94	7.76*	12.56	1.94	14.50
C	4.80	3.45	8.25	10.36	3.45	13.81*
C – B	–1.02	+1.51	+0.49	–2.20	+1.51	–0.69
	Saving on labour cost less than extra capital cost *therefore*		Increase in combined cost	Saving on labour cost greater than extra capital cost *therefore*		Decrease in combined cost

* Asterisk denotes minimum cost quantity-combination after neutral technological change.

(*a*) For A, B and C the respective unit costs are calculated by applying the appropriate factor prices to the relevant unit input requirements from rows (6), (3) and (1) respectively in Table 2.1 (taking row (1), third and fourth columns, because this gives the requisite capital–labour ratio of 8.63 with labour productivity of 500 bricks per labour-hour (= 1/0.00200 labour-hours per brick)).

Source: see text.

Therefore, the increased capital–labour ratio with its concomitant increased labour productivity is not economic.

Thus we can see that having a higher capital-intensity of production and higher labour productivity does not necessarily ensure a minimum unit cost of production. It is this scenario that would justify the complaint: 'That increase in labour productivity [from B to C] does not tell you very much; that plant has increased its labour productivity simply by installing excessive, unnecessarily large, amounts of fixed capital.'

In the second scenario, where the factor price ratio has changed to 5.18/ 0.20 = 25.9, B is not a minimum cost position and C is a minimum cost position. The arithmetic of this can be seen in Table 2.2's second scenario. Under the new factor price ratio, the C–B comparison shows that capital–labour substitution would lead to a saving in unit labour cost of $2.20 per thousand bricks and the extra unit capital cost would be only $1.51 per thousand bricks, so making for a net difference in unit labour and capital cost combined of –$0.69 per thousand bricks.

Therefore, under the changed factor price ratio the increased capital–labour ratio with its concomitant increased labour productivity *is* economic; the plant with the higher capital–labour ratio of $8.63 of capital installed per labour-hour used per annum would have the lowest achievable unit labour and capital cost combined.

The increased factor price ratio makes it necessary to have the higher capital–labour ratio in order to achieve minimum cost production. In this case the complaint about the installation of extra fixed capital would not be justified. Such capital–labour substitution in response to a change in the factor price ratio and undertaken to achieve the minimum unit cost position may be called 'mechanization'.

In either scenario we can decompose the increase in labour productivity into two components: the increase from A to B which is ascribed to the impact of technological change; and the increase from B to C which is ascribed to the impact of capital–labour substitution.

However, whenever labour productivity is increased through increased multifactor productivity, then both unit labour requirement and unit capital requirement will be reduced, and this will certainly reduce unit labour and capital cost combined. Thus, ultimately, comparisons among plants may be largely determined by what happens to multifactor productivity. Over a period of time, as plants with higher multifactor productivity, and consequently lower unit cost, replace plants with lower multifactor productivity, and consequently higher unit cost, an entire industry and indeed the economy itself may benefit from increases in multifactor productivity.

This we may now see by considering some data relating to manufacturing industry in the United States.

2.6 Multifactor productivity: United States manufacturing industry

Multifactor productivity can be measured as the ratio between the quantity of output per period, Q_t, and the combined quantities of factor inputs, where the combined input is measured as $L_t^a K_t^b$.

In practice, multifactor productivity is usually measured as the ratio between an index of the quantity of output per period and an index of the combined quantities of factor inputs, the index values being set equal to 100 in some base year.

When multifactor productivity is measured in this way, the crucial issue is understanding how to calculate an index of combined factor input. The index formula for the combined factor input which is most commonly used is known as the Törnqvist index, and this index will be fully explained.

When output is measured for a segment of the economy, rather than for a plant producing one type of product, output has to be measured as value added at constant prices. We shall denote value added at constant prices by the symbol V. But value added is generally taken in index form, so in most of what follows V should be taken to denote the index value of value added at constant prices. Likewise, the symbols L and K should respectively be taken when appropriate to denote index values of the quantity of labour input and capital installed.

We may illustrate how multifactor productivity may be measured by using some data for manufacturing industry in the United States taken from a study by the United States Department of Labor Bureau of Labor Statistics, *Trends in Multifactor Productivity, 1948–81* (Bulletin 2178, September 1983), hereafter referred to as Bulletin 2178. The index values are reported in Bulletin 2178 with 1977 = 100, but for convenience, I have altered this to 1948 = 100 by dividing all values by the index value for 1948. This leads to a few very small rounding discrepancies, but these need be of no concern.

Table 2.3 gives the data. A long period is covered, and this is useful in quantifying the important contribution of multifactor productivity increase to the growth in total output. Between 1948 and 1981, total annual output (value added at constant prices) of manufacturing industry in the United States increased nearly threefold, or by 3.3 per cent per annum compound.

The measurement of the total quantity of labour input is the total quantity of paid labour-hours worked per annum by all persons working for pay in manufacturing industry plus labour-hours worked by proprietors of enterprises and labour-hours worked by unpaid family workers. To be absolutely explicit, this index is based on the aggregate of labour-hours worked by *all* workers in

Table 2.3
Multifactor productivity change in United States manufacturing industry, 1948 to 1981

	Index values: 1948 = 100				
	(1)	(2)	(3)	(4)	(5) = *(1)/(4), subject to rounding error*[a]
Year, t	Value added at constant prices, V_t	Paid hours of all employees plus hours of proprietors and unpaid family workers, L_t	Flow of capital services used, K_t	Chained index of combined labour and capital inputs, $L_t^{\hat{a}} K_t^{\hat{b}}$	Multifactor productivity index, $Z_t' = V_t/(L_t^{\hat{a}} K_t^{\hat{b}})$
1948	100.00	100.00	100.00	100.00	100.00
1949	94.69	91.18	104.22	95.13	99.54
1950	107.82	98.49	107.39	101.26	106.48
1951	120.11	106.05	113.98	108.48	110.73
1952	124.30	107.56	122.43	112.09	110.90
1953	132.68	113.10	127.18	117.43	112.99
1954	123.18	103.40	130.61	111.15	110.83
1955	136.59	109.07	134.56	116.48	117.26
1956	137.43	110.71	140.37	119.15	115.34
1957	138.27	108.94	145.65	119.31	115.89
1958	126.26	100.00	148.28	112.87	111.86
1959	141.06	106.68	149.34	118.37	119.17
1960	141.62	106.30	151.72	118.68	119.33
1961	141.62	103.65	153.83	117.11	120.93
1962	153.91	107.81	156.20	120.88	127.33
1963	166.48	108.94	160.16	122.76	135.61
1964	178.49	111.34	164.64	125.59	142.12
1965	194.97	117.88	171.77	132.34	147.33
1966	209.78	125.69	182.59	140.97	148.81
1967	209.50	125.44	195.78	143.96	145.53
1968	220.95	127.71	206.33	148.19	149.09
1969	228.21	129.85	214.51	151.65	150.49
1970	215.08	122.54	221.37	146.94	146.38
1971	219.83	118.01	224.80	143.64	153.04
1972	240.78	123.17	227.97	148.67	161.96
1973	267.88	129.97	233.77	155.57	172.19
1974	256.70	127.46	243.27	155.10	165.51
1975	238.55	115.11	251.98	145.37	164.10
1976	261.45	120.78	256.99	151.33	172.76
1977	279.33	125.94	263.85	156.99	177.93

cont.

Table 2.3 (continued)
Multifactor productivity change in United States manufacturing industry, 1948 to 1981

	Index values: 1948 = 100				
	(1)	(2)	(3)	(4)	(5) = *(1)/(4), subject to rounding error*[(a)]
Year, *t*	Value added at constant prices, V_t	Paid hours of all employees plus hours of proprietors and unpaid family workers, L_t	Flow of capital services used, K_t	Chained index of combined labour and capital inputs, $L_t^{\hat{a}} K_t^{\hat{b}}$	Multifactor productivity index, $Z_t' = V_t / (L_t^{\hat{a}} K_t^{\hat{b}})$
1978	294.13	131.49	273.88	163.58	179.81
1979	302.23	134.13	287.07	168.29	179.59
1980	289.39	128.21	303.69	164.99	175.39
1981	295.81	127.58	319.53	166.41	177.77
Total increase 1948–81	*Just under threefold*	*Just over one-quarter*	*Just over threefold*	*Just two-thirds*	*Just over three-quarters*
Compound annual rate of increase, per cent per annum	*3.3*	*0.7*	*3.6*	*1.6*	*1.8*

(*a*) Calculated from data in source as: $((V_t/(L_t^{\hat{a}} K_t^{\hat{b}}))/(V_{1948}/(L_{1948}^{\hat{a}} K_{1948}^{\hat{b}}))) \times 100$. Accordingly, there may occasionally be very slight rounding discrepancies with results obtained by dividing rebased and rounded number in column (1) by rebased and rounded number in column (4).

Source: United States Department of Labor Bureau of Labor Statistics, *Trends in Multifactor Productivity, 1948–81*, Bulletin 2178, September 1983, table 10, p. 24; data rebased to 1948 = 100.

manufacturing industry during the relevant year; that is, the annual labour-hours worked by *each and every* worker in manufacturing industry are added together to give the required aggregate annual quantity of labour input. This aggregate is given not in original values (of so many millions of labour-hours) but in index form.

Between 1948 and 1981 this measure of the quantity of annual labour input increased by just over one-quarter, or by 0.7 per cent per annum. It is thus apparent that the quantity of annual labour input increased rather slowly, and very much more slowly than did output. This is, of course, the counterpart to labour productivity growth.

The growth of capital input, as measured by the Bureau of Labor Statistics, increased just over threefold, or by 3.6 per cent per annum. The quantity of capital input thus grew at about the same rate, or just a little faster, than did output. It is thus clear that the capital–labour ratio in manufacturing industry must have risen over this period.

This pattern of growth – slow growth of labour input with growth of capital input a little more than matching the growth of output – is reasonably typical of growth in manufacturing industry in most economies, and is also typical of long-run growth in most enterprises in manufacturing industry.

It is a pattern of growth which comes about because capital is a produced means of production and so an increase in the input of capital can readily be obtained, while the same is not true of labour input, especially if (the demographically limited supply of) workers have a predilection for working fewer hours per week or a resistance to increasing weekly hours of work. Consequently, labour productivity can be raised by capital–labour substitution as well as by technological change.

Using these data, we may measure the extent of multifactor productivity change, which, over such a long period, is largely determined by technological change (although the effect of economies of scale is likely also to have been of significance).

We can do this if we can measure the *combined* input of labour and capital or 'combined factor input'. How may this be done?

To ask this question is, implicitly, to ask about the form of the production function, because the form of the factor input frontier production function must specify the way in which factor inputs are combined in order to produce output. We shall continue to use the simple production function so far used in this book, namely $Q = ZL^aK^b$, although Bulletin 2178 used a more generalized form of function. However, the simple production function used here suffices to explain the basic methodology involved in measuring combined factor input.

The details showing how the index of combined factor input was calculated are given in Table 2.4. These calculations are now explained step by step.

As explained, the parameter a is the (partial) elasticity of output with respect to the labour input; that is, a may be understood as the proportionate response of annual output with respect to a proportionate change in the annual input of

Table 2.4
Calculating the index of combined factor input as a chained index, United States manufacturing industry

	Year, t				
	0, 1948	1, 1949	2, 1950	3, 1951	4, 1952
(1) Share of labour in value added, $(W/V)_t$	0.681	0.670	0.656	0.661	0.683
(2) Two-year moving average labour share, $(1/2)((W/V)_{t-1} + (W/V)_t) = \hat{a}$	–	0.6755	0.6630	0.6585	0.6720
(3) Two-year moving average capital share, $1 - \hat{a} = \hat{b}$	–	0.3245	0.3370	0.3415	0.3280
(4) Proportionate ratio of labour input, L_t/L_{t-1}	–	0.9118	1.0802	1.0768	1.0142
(5) Proportionate ratio of capital input, K_t/K_{t-1}	–	1.0422	1.0304	1.0614	1.0741
(6) $(L_t/L_{t-1})^{\hat{a}}$	–	0.9395	1.0525	1.0499	1.0095
(7) $(K_t/K_{t-1})^{\hat{b}}$	–	1.0135	1.0101	1.0206	1.0237
(8) Proportionate ratio of combined factor input between successive years, (6) × (7)	–	0.9522	1.0632	1.0715	1.0335
(9) Index of combined factor input[(a)]	100	95.22	101.24	108.48	112.11

(*a*) Using unrounded data and hence differing very slightly from figures in column (4) of Table 2.3.

Source: Table 2.3 (for rows (4) and (5)) and US Department of Labor Bureau of Labor Statistics, *Trends in Multifactor Productivity, 1948–81*, Bulletin 2178, September 1983, table C-5, p. 54 (for row (1)). See also Bulletin 2178, Appendix A, The Multifactor Productivity Model, especially p. 34, equations A.4, A.5 and A.6 for the Bureau of Labor Statistics (BLS) method of calculating the index of combined factor input ('the aggregate index of [factor] inputs') and the first paragraph of the second column of print on p. 34 for a brief justification of the BLS method.

labour (when the input of capital is held constant). Likewise, *mutatis mutandis*, for the input of capital and the parameter b. Consequently, the value of a and the value of b are both important in determining the behaviour of producers because each indicates how output would vary with a variation in the respective factor input.

The Bureau of Labor Statistics estimated the value of a as the share of value added being paid in wages and salaries to workers. The argument in support of this method of estimating a value for a may be explained in the following six steps. For the duration of this explanation, it is necessary to use measures of V_t, L_t, and K_t taken in absolute (original) values (not in index values).

First, it is necessary for this explanation to redefine the production function in terms of value added at constant prices as a measure of output, measured in dollar terms; likewise, labour input is taken in labour-hours and capital stock is measured in dollars at the same constant prices as used for value added. We then have the production function:

$$V_t \quad = \quad Z'_t L_t^a K_t^b$$

where we now use Z' to denote the adjustment factor (because of the different measure of output being used). This adjustment factor still performs the same function as in the production function for the (physical) quantity of output Q_t.

Second, we note that the increase in annual value added with respect to an extra unit of labour input (labour-hours) per annum is:

$$\mathrm{d}V_t/\mathrm{d}L_t \quad = \quad a\, Z'_t L_t^{a-1} K_t^b$$

This is the marginal value product of labour. In other words, it is the increase in the money value of output (value added in dollar terms at constant prices) with respect to an additional unit by physical quantity of labour input, measured here in labour-hours. It is, effectively, the extra value added produced per annum by an increase of 1 labour-hour in the annual quantity of labour input.

Note that the marginal value product of labour is not independent of the total quantity of labour input being used. Specifically, under the restriction $a + b = 1$, the exponent on the labour input in the marginal value product equation, $a - 1$, is *negative*, so that the greater is L_t, the smaller is the marginal value product of labour.

Third, and this is the critical step of the explanation, if, in general, a labour-hour is paid according to its marginal value product, then $\mathrm{d}V_t/\mathrm{d}L_t$ should equate to (or at least should closely approximate to) the wage rate paid per labour-hour. This measure of the wage rate should also be in constant price terms, to conform with value added measured in constant price terms. We shall denote this (constant price) wage rate per labour-hour as w_t. This distinguishes the constant price or 'real' wage rate as used here from the symbol p_L previously used to denote the price of a unit of labour.

Why should the wage rate approximate to the marginal value product of labour? The answer is that if the marginal value product of labour were *greater* than the wage rate, then by hiring an extra unit of labour the enterprise would obtain a greater increase in the money value of output (value added) than it would incur additional cost of labour and so there would be, for the enterprise, an increase in the annual gross operating surplus from hiring such an extra unit of labour. With no additional capital, and hence with no additional annual depreciation, the increase in the annual gross operating surplus means that the annual net operating surplus, or profit, will also increase, as will the rate of profit per annum.

Accordingly, on this argument, the profit-seeking enterprise is likely to hire this extra unit of labour.

Note that, according to the production function, this will diminish the marginal value product of a further unit of labour. This is essential to the argument. Obtaining an extra unit of labour may also, by the working of the labour market, require some increase in the wage rate paid to attract that extra unit of labour (but this is not essential to the argument).

If the marginal value product of labour remains greater than the wage rate, the process of hiring extra labour will continue, with the certain effect (according to the production function) of lowering the marginal value product of labour and with the likely effect (according to the labour market) of raising the wage rate.

Eventually, the marginal value product of labour will equal (or approximate to) the wage rate and at this point the hiring of additional units of labour will cease.

Conversely, if the marginal value product of labour were less than the wage rate, then by reducing annual labour input by one unit of physical quantity, the enterprise would lose some money value of output (annual value added), but this loss will be less than the annual cost saving from the reduced labour input, and so there is, for the enterprise, an increase in the annual gross operating surplus from reducing its labour input with a consequent increase in the annual net operating surplus and in the rate of profit per annum.

Accordingly, on this argument, the enterprise is likely to reduce its labour input. Note that according to the production function this reduction in labour input will raise the marginal value product of labour. Reducing labour input may also, by the working of the labour market, lead to some decrease in the wage rate paid (but this is not essential to the argument).

Eventually, the marginal value product of labour will equal (or approximate to) the wage rate, and at this point the reduction in the labour input will cease.

If we accept this argument about the equality (or approximation) between the marginal value product of labour and the wage rate, then, fourth, we can rewrite the preceding equation for the marginal value product of labour, substituting, on the left-hand side, w_t for dV_t/dL_t:

$$w_t \quad = \quad a\,Z'_t L_t^{a-1} K_t^b$$

Fifth, if we multiply this equation through by L_t, the total quantity of labour-hours used per annum, we obtain an expression for the total amount paid in wages each year, generally referred to as the 'wage bill' or synonymously as 'labour remuneration', and which we shall denote as W_t where $W_t \equiv w_t L_t$:

$$\begin{aligned} W_t &= a\,Z'_t L_t^{a-1} K_t^b L_t \\ &= a\,Z'_t L_t^a K_t^b \\ &= a\,V_t \end{aligned}$$

Sixth and finally, we can then estimate the numerical value of the parameter a, or the partial elasticity of output with respect to the labour input, as the ratio between the annual wage bill and annual value added:

$$a \quad = \quad W_t/V_t$$

Conversely, in a constant returns to scale production function, the value of the parameter b may be estimated by taking the share of the gross operating surplus in value added. Or, equivalently, the value of b may be measured as $1 - a$.

This is because the gross operating surplus is simply what is left from value added after the total amount of wages has been deducted from value added. If we let G denote the annual gross operating surplus (measured in dollar values at constant prices), then:

$$G \quad = \quad V - W$$

so that:

$$G/V = V/V - W/V$$
$$= 1 - a$$

In estimating a as W/V, we have explained that both W and V should theoretically be understood as being measured in constant price terms, and this must always be the case whenever W or V are taken in isolation. However, from the point of view of the *ratio* W/V it does not make a material or practical difference if the ratio between the absolute values is measured in terms of current price variables. This is because the numerator and the denominator in the current price ratio would each be divided by the same output price index to give a ratio in terms of constant price variables.

Accordingly, it is also possible to estimate the value of a using the ratio between current price variables. The Bureau of Labor Statistics estimated values of a as the share of current price labour remuneration in current price value added (Bulletin 2178, pp. 33–4). Table 2.4 shows in the first row the share of labour remuneration in value added (W_t/V_t) for a sample of the first five years. It can be seen that there are slight variations, year to year, in the share of labour.

However, the parameter a in the production function is a constant. The question then arises how best to use the series of observations W_t/V_t in the place of the parameter a. There are various possibilities.

It is not possible to use each year's observation on the labour share as an estimate of a because we would then have simultaneous variation, year to year, in the factor inputs L and K and in the parameters a and b. Consequently we cannot on this method of using W/V calculate an index of combined factor input: the index of combined factor input is supposed to be an index relating to changes in L and K only, so it cannot also include changes in a and b.

One possibility is simply to use the first year's observation, W_0/V_0 as an estimate of a, which is then taken to be constant for the whole period. The problem with this is that, if the partial elasticity of output with respect to the labour input is actually changing, and if producers are (gradually) adjusting to this change, then the first observation does not represent the economic reality facing producers over the whole period.

Here it is important to bear in mind the argument that the capital–labour ratio at the minimum unit cost position, $(K/L)^*$, is determined by, and is equal to, the expression $(p_L/p_K)/(a/b)$, so that $(K/L)^*$ changes if a and b change, and producers will react accordingly.

Another possibility is to average W_t/V_t over all observations and to use this average as the estimate of a, which is then taken to be constant for the whole period. The problem with this is that, if there is a trend in W_t/V_t then in the early years producers are assumed somehow to have already reacted to the average value of this parameter in advance of its change rather than adjusting to the situation confronting them at the time of production.

A theoretically and practically convenient compromise possibility is to have a constant parameter for each of the year-to-year movements in each of L and K but to permit the parameter to vary *after* each year-to-year movement, and then to link the year-to-year movements by the process known as 'chaining'. This procedure acknowledges theoretically that producers may be gradually adjusting to alterations in the production function parameters, but it still permits practically a measure of the movement in the combined factor input. This is the method which was adopted in the Bureau of Labor Statistics measure of combined factor input (Bulletin 2178, p. 34). Let us see how this method works.

The first row of Table 2.4 gives the share, in each year, of labour remuneration in value added, W_t/V_t. The next step is to calculate the two-year moving average of this share, and the results of these calculations are given in the second row of Table 2.4. To illustrate, the two-year moving average for 1949 is calculated as (1/2)(0.681 + 0.670) = 0.6755.

This calculation is repeated for each pair of years. The two-year moving average for 1950 is calculated as (1/2)(0.670 + 0.656) = 0.6630. And so on.

Each of these averages may be called a 'two-year moving average'; 'two-year' obviously enough because the average is based on only two years, and 'moving average' because the average 'moves', or changes, successively over time.

We shall designate each pair of successive years as $t-1$ and t respectively; the use of '−1' indicating that we are referring to a year t and its *immediately preceding* year. Each of these two-year moving averages may then be used as an estimate of the production function parameter a, which is held constant between year $t-1$ and year t. We shall designate this constant estimate as $\hat{a}$. Note that we shall not put a time subscript on $\hat{a}$ because, for each *pair* of years, $\hat{a}$ is constant, even though $\hat{a}$ varies as between *pairs* of years, as we shall show. However, when we use the symbol $\hat{a}$ you should remember that it relates only to the pair of successive years $t-1$ and t, and that when t changes, so too will $\hat{a}$.

Once we have calculated the two-year moving average labour share, the two-year moving average capital share (the share of the gross operating surplus in value added) may simply be calculated as $1-\hat{a}$. The two-year moving average capital share may likewise be designated $\hat{b}$, where $\hat{b} = 1 - \hat{a}$.

We now explain how $\hat{a}$ and $\hat{b}$ may be used in calculating the index of combined factor input. For this argument, and in conformity with the data given in Table 2.3, we revert to index values of all variables.

For t and $t-1$ we have the respective value added production functions:

$$V_t \quad = \quad Z'_t L_t^{\hat{a}} K_t^{\hat{b}}$$

and:

$$V_{t-1} \quad = \quad Z'_{t-1} L_{t-1}^{\hat{a}} K_{t-1}^{\hat{b}}$$

According to the production function using the two-year moving average estimates of the output elasticities, the combined factor input for year t is $L_t^{\hat{a}} K_t^{\hat{b}}$, and the combined factor input for year $t-1$ is $L_{t-1}^{\hat{a}} K_{t-1}^{\hat{b}}$, where $\hat{a}$ has the same value in each expression and $\hat{b}$ has the same value in each expression.

Accordingly, the proportionate change, in ratio form, between year $t-1$ and year t in the combined factor input may be calculated as:

$$\frac{L_t^{\hat{a}} K_t^{\hat{b}}}{L_{t-1}^{\hat{a}} K_{t-1}^{\hat{b}}} = \frac{\text{Index of combined factor input in } t}{\text{Index of combined factor input in } t-1}$$

$$= \begin{array}{l}\text{Proportionate change, in ratio form,} \\ \text{in index of combined factor input between} \\ t-1 \text{ and } t\end{array}$$

It is necessary to note, explicitly, that this ratio, taken as a single entity, is the proportionate change in the index of combined factor input.

Because the same (identical) values of $\hat{a}$ and $\hat{b}$ appear in the numerator and the denominator, the ratio may be conveniently rewritten as:

$$(L_t/L_{t-1})^{\hat{a}} \times (K_t/K_{t-1})^{\hat{b}} = \begin{array}{l}\text{Proportionate change, in ratio form,} \\ \text{in index of combined factor input between} \\ t-1 \text{ and } t\end{array}$$

On the left-hand side this is a more convenient formula because we have the proportionate change in labour input between $t-1$ and t and the proportionate change in capital input between $t-1$ and t. Each of these proportionate changes is raised to the power, or exponentiated by, $\hat{a}$ and $\hat{b}$

respectively, and (this is the important point) the product of the exponentiated proportionate changes is equal to the proportionate change in the combined factor input.

This demonstrates that the proportionate change in the combined factor input can be calculated, and can be thought of, as a sort of weighted 'average' of the proportionate change in each of the labour input and the capital input, where the weights are the exponents $\hat{a}$ and $\hat{b}$ respectively.

Thus the next step in Table 2.4 is to calculate, for each pair of years, the proportionate change, in ratio form, of each of the quantity of labour and the quantity of capital input. Accordingly, rows (4) and (5) of Table 2.4 present the year-to-year change, in ratio form, of columns (2) and (3) respectively of Table 2.3. (The calculations in Tables 2.3 and 2.4 use unrounded values of each variable, so do not be worried by minor rounding discrepancies when you repeat these calculations on your calculator.)

To illustrate: for 1949 (or t) the proportionate change over 1948 (or t–1) in the labour input is calculated from column (2) of Table 2.3 as 91.18/100 = 0.9118; for 1950, the proportionate change, *over 1949*, in the labour input is calculated as 98.49/91.18 = 1.0802.

Likewise for the capital input in column (3) of Table 2.3 with results given in row (5) of Table 2.4.

The next step is to exponentiate, or 'weight', each of these proportionate changes in factor inputs by its appropriate two-year moving average factor share: $\hat{a}$ for the proportionate change in labour and $\hat{b}$ for the proportionate change in capital input. These calculations are given in rows (6) and (7) of Table 2.4 respectively.

To illustrate: for 1949, the 'weighted' change in ratio form, over 1948, of the labour input is $0.9118^{0.6755} = 0.9395$, and the 'weighted' change in ratio form, over 1948, of the capital input is $1.0422^{0.3245} = 1.0135$. For 1950, the 'weighted' change in ratio form, *over 1949*, of the labour input is $1.0802^{0.6630} = 1.0525$, and the 'weighted' change in ratio form, *over 1949*, of the capital input is $1.0304^{0.3370} = 1.0101$.

The next step is to multiply each year's exponentiated, or 'weighted', change for the labour input by the exponentiated, or 'weighted', change for the capital input. The result of this multiplication is to give the proportionate change in the combined factor input, this change being in ratio form *between each year and its immediately preceding year*.

To illustrate: between 1948 and 1949 the combined factor input changed by the ratio 0.9522, calculated as 0.9395 × 1.0135. Between 1949 and 1950 the combined factor input changed by the ratio 1.0632, calculated as 1.0525 × 1.0101. (The calculations in Tables 2.3 and 2.4 used unrounded values of each variable, so do not be worried by minor rounding discrepancies when you repeat these calculations on your calculator.)

The final step in calculating the index of combined factor input is to use each of these year-to-year proportionate changes in the combined factor input to calculate a 'chain' index of combined factor input or 'multifactor input'. The calculation is done as follows.

First, the index of combined factor input can be set equal to 100 in the base year 1948. This is because each of the factor inputs has an index value of 100 in 1948 and because $a + b = 1$ so that for 1948 the combined factor input is: $100^a \times 100^b = 100^{a+b} = 100^1 = 100$. This result is given in row (9) of Table 2.4.

We then know, or have estimated, that between 1948 and 1949 the combined factor input changed by the ratio 0.9522 (row (8) of Table 2.4), so it follows that, second, in 1949 the combined factor input index had a value of 100 × 0.9522 = 95.22. This result is given in row (9) of Table 2.4.

Third, we know, or have estimated, that between 1949 and 1950 the combined factor input changed by the ratio 1.0632 (row (8) of Table 2.4), so it follows that in 1950 the combined factor input index had a value of 95.22 × 1.0632 = 101.24. This result is given in row (9) of Table 2.4.

Fourth, we know, or have estimated, that between 1950 and 1951 the combined factor input changed by the ratio 1.0715, so it follows that in 1951 the combined factor input index had a value of 101.24 × 1.0715 = 108.48. This result is given in row (9) of Table 2.4. And so on.

We can thus see that the index of combined factor input is calculated according to a 'chain' process as follows (bold print indicates successive values of the index of the combined factor input as these emerge from the process of 'chaining' (row (9)) and italics indicates the successive 'links' in the chain (row (8))):

				Year
100				$t = 0$
100 × *0.9522* = **95.22**				$t = 1$
	95.22 × *1.0632* = **101.24**			$t = 2$
		101.24 × *1.0715* = **108.48**		$t = 3$

You can now see why this process of calculating the value of an index from each of successive proportionate links is known as chaining. The index in row (9) of Table 2.4 can accordingly be referred to as a 'chained index'.

The index can also be called a 'Törnqvist index' after the statistician who was among the first to use the general form of this index. Törnqvist used moving average weights of shares in consumers' expenditure and a chain method to calculate a consumer price index. He argued that the advantage of weights which were 'variable in principle', together with the chain method, was that it provided 'the possibility, thanks to the weighting system being changeable, of an elastic adjustment to the alterations in the composition of consumption that occur in the course of time' (Leo Törnqvist, 'The Bank of Finland's Consumption Price Index', Bank of Finland, *Monthly Bulletin*, 1936, No. 10, pp. 28 and 32). This is exactly the same reason (in another context) why the calculation of the index of combined factor input uses a two-year moving average of weights together with the method of chaining.

The chained index of the combined factor input is given in column (4) of Table 2.3 (small differences with row (9) of Table 2.4 are due to rounding discrepancies).

For manufacturing industry in the United States, combined factor input increased by two-thirds over the whole period 1948 to 1981, or by about 1.6 per cent per annum. This rate of increase is the result of the combination of the slow rate of growth of labour input and of the faster rate of growth of capital input.

Having calculated the chained index of combined factor input, the index of multifactor productivity, Z'_t, with a base value of 100, can be calculated for each year t as the ratio between the index of output and the chained index of combined factor input:

$$Z'_t = (V_t/(L_t^{\hat{a}} K_t^{\hat{b}}))\ 100 \qquad \textit{Index of multifactor productivity}$$

To illustrate for 1949 using the data given in Table 2.3:

$$Z'_{1949} = \left(\frac{\text{Index of value added at constant prices in 1949}}{\text{Chained index of combined factor input in 1949}}\right)100$$

$$= (94.69/95.13)\ 100$$

$$= 0.9954 \times 100$$

$$= 99.54$$

and for 1950:

$$Z'_{1950} = (107.82/101.26)\,100$$
$$= 1.0648 \times 100$$
$$= 106.48$$

And so on.

The results for the index of multifactor productivity show that multifactor productivity increased by just over three-quarters during the period 1948 to 1981, or by about 1.8 per cent per annum.

We can thus say that just under one-half of the trend rate of growth of output of manufacturing industry of 3.3 per cent per annum resulted from the growth of combined factor input (1.6 per cent per annum/3.3 per cent per annum), and that just over one-half of the growth of output of manufacturing industry resulted from the increase in multifactor productivity (1.8 per cent per annum/3.3 per cent per annum).

The results show that the increase in multifactor productivity was quantitatively very important in contributing to the growth of output in manufacturing industry in the United States.

Conversely, the growth rate of output was considerably greater than can be accounted for simply by the increase in combined factor input. The proportionate increase in multifactor productivity over the entire period means that in 1981 United States manufacturing industry produced 77.77 per cent more from the same quantities of factor inputs than it did in 1948.

We can see from the data in Table 2.3 that capital input grew by more over the entire period than did the labour input; in other words, the capital–labour ratio increased over this period. In fact, the increase in the capital–labour ratio must have been in the ratio 319.53/127.58 = 2.50. In 1981, each hour of annual labour input was 'supported' by 2.5 times the amount of fixed capital input than in 1948.

This increase in the capital-intensity of production would have raised labour productivity considerably. The increase in multifactor productivity by a factor of 1.7777 then measures the (hypothetical) increase in labour productivity which would have occurred had the increase in the capital–labour ratio not happened. It is the increase in labour productivity 'adjusted for the influence of' (or 'subtracting the effects of') the increase of capital per labour-hour.

Many other studies have shown a similar quantitative importance to economic growth of the increase in multifactor productivity, which usually contributes about, and sometimes over, one-half of the growth in output. (A selected sample of these is: Edward F. Denison, *Why Growth Rates Differ:*

Postwar Experience in Nine Western Countries (The Brookings Institution, 1967), chapter 21; Edward F. Denison, *Trends in American Economic Growth, 1929–1982* (The Brookings Institution, 1985), p. 22 and tables 7-1 and 8-1, pp. 107 and 111; R.C.O. Matthews, C. H. Feinstein and J.C. Odling-Smee, *British Economic Growth 1856–1973* (Oxford University Press, 1982), table 8.3, pp. 228–9; Organization for Economic Co-operation and Development, *OECD Economic Studies* (No. 10, Spring 1988), table 1, pp. 14–15; Australian Bureau of Statistics, C. Aspden, *Occasional Paper: Estimates of Multifactor Productivity, Australia* (ABS Catalogue No. 5233.0, August 1990), tables 9.1 and 9.2, pp. 13 and 14.)

If we accept the argument that in the long run the increase in multifactor productivity is largely determined by technological change, then all this demonstrates quantitatively the long-run contribution of technological change to economic growth.

3 Non-neutral Technological Change in Theory

3.1 Defining non-neutral technological change

Neutral technological change means that, at the minimum cost position, the capital–labour ratio remains constant (providing the factor price ratio also remains constant). Although helpful as an introduction to the basic economics of technological change, this analysis of technological change restricted to a constant capital–labour ratio is likely to prove somewhat unrealistic because technological change may well require, or lead to, the substitution of capital for labour *as an integral part of the change.*

Accordingly, we need an analysis of technological change where technological change leads to a change in the capital–labour ratio at the minimum cost position (with a constant factor price ratio). This is a less restricted analysis and accords with what happens in the real world.

The capital–labour ratio at the minimum cost position, $(K/L)^*$, is determined by the factor price ratio, (p_L/p_K), divided by the ratio of the exponents to the production function, (a/b); that is: $(K/L)^* = (p_L/p_K)/(a/b)$ – see Chapter 2.

Accordingly, change in the capital–labour ratio at the minimum cost position, or capital–labour substitution, can be incorporated into the analysis of technological change by letting the technological change alter the exponents in the frontier production function (in addition to increasing the adjustment factor in the production function). Throughout we use a production function which has the restriction of constant returns to scale where $a + b = 1$, so that an alteration in the exponents necessarily leads to a change in the ratio of the exponents, a/b.

If the ratio a/b is altered by technological change then the capital–labour ratio, at the minimum cost position, must be affected by technological change (if the factor price ratio remains constant). This feature leads to the technological change being referred to as 'non-neutral'. Taking the minimum cost positions, the direction of the non-neutrality may be identified according to whether the capital–labour ratio increases or decreases by comparison with the minimum cost capital–labour ratio before technological change.

If the capital–labour ratio increases (that is, if capital is substituted for labour), then the technological change is called 'capital-using' or, conversely, 'labour-saving'. If the capital–labour ratio decreases (that is, if labour is substituted for capital), then the technological change is called 'capital-saving' or 'labour-using'. We have the following synopsis of possibilities in the presence of a constant factor price ratio:

If the ratio of exponents, a/b:	*Then the capital–labour ratio at minimum cost combination, $(K/L)^*$:*	*Consequently, non-neutrality is referred to as:*
Falls	Increases	Capital-using/labour-saving non-neutral technological change
Rises	Decreases	Capital-saving/labour-using non-neutral technological change

In all of what follows we concentrate on capital-using/labour-saving non-neutral technological change because such change enhances the increase in labour productivity and so has important implications for growth in standards of living.

The analysis of non-neutral technological change has to deal with two separate but related issues. The first issue is the comparison of the situation before non-neutral technological change with the situation after non-neutral technological change. This I call the 'before and after comparison'.

The second issue is the comparison of the situation after *neutral* technological change with the situation after non-neutral technological change. This I call the 'neutral/non-neutral comparison'.

Because we have two separate comparisons to make, we require some additional notation, not previously used, in order to distinguish between the two types of technological change. We have the following synopsis:

Plant of vintage t	*Plants of vintage t + 1*
Plant I, before technological change, minimum cost position designated	Plant II, after *neutral* technological change, minimum cost position designated

by superscript *, vintage designated by subscript t	by superscript *, vintage designated by subscript $t + 1$
	Plant III, after capital-using/labour-saving *non-neutral* technological change designated by superscript ′, minimum cost position designated by superscript *, vintage designated by subscript $t + 1$

The before and after comparison is between Plant I and Plant III; the neutral/non-neutral comparison is between Plant II and Plant III.

Under constant returns to scale, if the value of the partial elasticity of output with respect to capital, b, increases and conversely the value of the partial elasticity of output with respect to labour, a, decreases, then the capital–labour ratio at the minimum cost position increases, and this is the defining criterion of capital-using/labour-saving non-neutral technological change: that is, at the same factor price ratio: $(K/L)'^{*}_{t+1} > (K/L)^{*}_{t}$.

Under a neutral technological change, the capital–labour ratio (at the minimum cost position) remains the same as the capital–labour ratio before the neutral technological change. Consequently (and all at minimum cost positions), the capital–labour ratio under capital-using/labour-saving non-neutral technological change is greater than the capital–labour ratio under neutral technological change. That is, at the same factor price ratio: $(K/L)'^{*}_{t+1} > (K/L)^{*}_{t+1}$.

This means that, by comparison with neutral technological change, capital-using/labour-saving non-neutral technological change involves capital–labour substitution or 'mechanization'.

The question of interest to the neutral/non-neutral comparison is whether technological change which requires greater mechanization will or will not give an advantage over technological change which does not require greater mechanization. In other words: Is it likely that the historical and present-day nature of technological change would be deliberately non-neutral and capital-using/labour-saving because greater cost reductions can thereby be obtained?

Having put the question in this way, you can see that the neutral/non-neutral comparison is important to understanding not only our economic history but also our economic future.

However, the before and after comparison must not be neglected because this comparison shows what happens to labour productivity and to unit cost, and this is a matter of great practical importance to plant managers and production engineers.

3.2 Labour productivity: before and after comparison

Comparisons must be made with respect to minimum cost positions, otherwise we shall not be comparing like with like. This is only common sense and will henceforth be taken as understood. In each expression the minimum cost position is signified by the use of an asterisk superscript, *.

From the labour productivity form of the production function before technological change, labour productivity in a plant of vintage t (Plant I) is:

$$(Q/L)^*_t = Z_t (K/L)^{*b}_t$$

Non-neutral technological change, for a plant of vintage $t + 1$, is signified by adding the superscript prime, $'$, to all relevant variables and parameters, as well as adding the subscript $t + 1$ to all variables. (But note that because the same value of Z_{t+1} is assumed for both neutral and non-neutral technological change, the superscript prime is not attached to Z; the reasons for this assumption are both to build on the example used in Chapter 2 and also to facilitate the requisite comparisons.)

Accordingly, labour productivity after non-neutral technological change (Plant III) is:

$$(Q/L)'^*_{t+1} = Z_{t+1} (K/L)'^{*\,b'}_{t+1}$$

Consequently, the overall proportionate impact on labour productivity of non-neutral technological change can be measured (in ratio form) as:

$$(Q/L)'^*_{t+1}/(Q/L)^*_t = \frac{Z_{t+1}}{Z_t} \frac{(K/L)'^{*\,b'}_{t+1}}{(K/L)^{*\,b}_t}$$

We shall refer to the ratio Z_{t+1}/Z_t as the first component of non-neutral technological change.

The last expression on the right-hand side contains a mixture of two differences: the difference in the capital–labour ratio – the difference between $(K/L)^*_t$ and $(K/L)'^*_{t+1}$ – and the difference in the partial elasticity of output with respect to the capital input – the difference between b and b'. We may conveniently analyse this mixed expression into two (further) distinct components as follows.

Multiply the right-hand side by $[(K/L)^{*\,b'}_t]/[(K/L)^{*\,b'}_t] = 1$, and then rearrange the resulting expression to get three (distinct) components involved in any non-neutral technological change (under the condition of constant returns to scale):

$$\frac{(Q/L)'^*_{t+1}}{(Q/L)^*_t} = \frac{Z_{t+1}}{Z_t} \times \frac{(K/L)'^{*\,b'}_{t+1}}{(K/L)^{*\,b}_t} \times \frac{(K/L)^{*\,b'}_t}{(K/L)^{*\,b'}_t}$$

$$= \frac{Z_{t+1}}{Z_t} \times (K/L)^{*\,b'-b}_t \times \left(\frac{(K/L)'^*_{t+1}}{(K/L)^*_t}\right)^{b'}$$

(1) Proportionate impact of neutral shift in production function

(2) Proportionate impact of exponent change on shift in production function

(3) Proportionate impact of non-neutral movement along new production function

(4) [(1) and (2)] Proportionate impact of non-neutral shift in production function

(5) [(4) and (3)] Proportionate impact of capital-using/labour-saving non-neutral technological change

This formula (that is, everything on the right-hand side of the equals sign) I call the *analytic formula for labour productivity change*, or the 'productivity formula' for short.

The analytic formula for labour productivity change is annotated with respect to each of its three distinct components, namely (1), (2) and (3); the composite item (4) – the product of components (1) and (2) – and the composite item (5) – the product of components (1), (2) and (3) – will be referred to as 'composite items' of the productivity formula.

Each of the three components is expressed in terms of the proportionate impact, measured in ratio form, on labour productivity.

The first component measures the impact of a neutral shift in the production function. This is one analytically distinct component of non-neutral technological change, and is simply measured by the proportionate change in the adjustment factor, Z.

(It is not absolutely necessary, in non-neutral technological change, that such a shift should occur. In other words, the adjustment factor Z could remain constant, but this would be simply a special case, and in what follows we shall take the general case where there is an increase in Z of the sort described and analysed previously for the case of neutral technological change.)

Before continuing, we digress for two paragraphs to note the following. In the case of *neutral* technological change (and under the restriction $a + b = 1$), $b' - b = 0$ (in other words, b does not change), so component (2) in the analytic formula would be equal to 1 (because any number raised to the power zero is equal to 1). Furthermore, under neutral technological change the capital–labour ratio remains constant so $((K/L)'^*_{t+1})/((K/L)^*_t) = 1$, and so component (3) in the analytic formula would be equal to 1 (because 1 raised to any power is equal to 1).

Consequently, for the specific case of neutral technological change, the impact of the technological change on labour productivity (at the minimum cost position) can be measured simply by the ratio Z_{t+1}/Z_t; that is, by component (1). This we have already seen in Chapter 2.

But in the case of non-neutral technological change there are three distinct components of the impact of the change on labour productivity (at the minimum cost position).

In the first place there is the impact of the increase in the adjustment factor from Z_t to Z_{t+1}. This component I call the neutral shift in the production function. In economic terms we may understand this component as follows. Technological change generally leads to an increase in output per unit of combined factor input – that is, to an increase in multifactor productivity.

This means that with the same total annual quantity of labour input and the same quantity of capital installed a greater annual quantity of output can be produced. The capital installed will not, of course, be of the same sort: it will be of a different sort in order to embody the new technology. But the same amount of money (*at constant prices*) will be spent on installing the new technology's capital equipment as would be (or would have been) spent on installing the old technology's capital equipment. The type of labour-hours may or may not change, but the same annual quantity of labour-hours can be used to produce, in conjunction with the same quantity of capital installed, a greater annual quantity of output.

Equivalently, technological change leads to an increase in labour productivity *at the same capital–labour ratio*. To paraphrase the words of the Bureau of Labor Statistics' *Trends in Multifactor Productivity, 1948–81*, Bulletin 2178: the change in multifactor productivity measures the change in labour productivity adjusted for the effect of the change in the capital–labour ratio. It is this component of the technological change which is measured by component (1). This component occurs under both neutral and non-neutral technological change.

With non-neutral technological change there are then two further distinct components to be taken into account: the impact of the change in the exponents of the production function; and the impact of the change in the capital–labour ratio. In the second place, according to the productivity formula, we may measure the impact on labour productivity of the further shift in the production function by component (2); namely, $(K/L)^{*\,b'-b}_{t}$.

This measures the impact of the change in the partial elasticity of output with respect to the capital input, and the impact is most appropriately measured at the initial capital–labour ratio *before* the technological change. Component (2) measures what I call the impact of the exponent change on the shift in the production function.

In the case of capital-using/labour-saving non-neutral technological change the ratio a/b falls to a'/b', so (under the restriction of constant returns to scale with the sum of the exponents being 1 in each case) b' must be greater than b and thus $b' - b$ must be positive. Consequently, for any value of the capital–labour ratio greater than 1, component (2) will be greater than 1. This means that component (2) measures a further increase in labour productivity. In other words, taken at the initial capital–labour ratio before technological change, the production function will have a *shift increase* additional to that given by the (neutral) increase in the adjustment factor, Z. This additional shift is a *skew* shift, because for any value of $b' - b$, the additional shift in the production function becomes greater the greater is the initial value of K/L. (To illustrate arithmetically, if $b' - b = 0.30 - 0.25 = 0.05$, then for $K/L = 4$ the additional shift increase is in the ratio $4^{0.05} = 1.072$, while for $K/L = 7$ the additional shift increase is in the ratio $7^{0.05} = 1.102$.)

Why does this additional and skew shift in the production function occur? What is its economic meaning?

In capital-using/labour-saving non-neutral technological change, the partial elasticity of output with respect to the capital input increases. This is the economic meaning of the change in *b*. The increased responsiveness of output to capital installed means that for any given proportionate increase in the quantity of capital installed, the proportionate gains in terms of extra output will be greater.

The result of this increase in *b* is that the plant with the new technology can produce more with the same quantity of factor inputs (regardless of the change in *Z*, or even if *Z* were not to increase). In this regard the impact measured by component (2) is indistinguishable in practice from the impact of the change in *Z* measured by component (1). Accordingly, it is often convenient to combine component (2) with component (1) into composite item (4), which is the entire shift in the production function resulting from non-neutral technological change. (The extent of this shift should be measured at the initial capital–labour ratio from which the change starts.)

In the analysis of technological change in the real world, it is this entire non-neutral shift in the production function which is of interest because it means that simply by using the same annual quantity of labour input and by spending the same amount of money (at constant prices) on installing capital (albeit a different sort of capital embodying the technological change), a plant can produce a greater annual quantity of output, so leading to greater productivity of each factor input. This increase in labour productivity is measured by composite item (4).

All this is most important for growth in the economy as a whole. This is because it is always possible for an economy to increase the quantity of capital installed in a way that it is not possible to increase the quantity of labour. Therefore an increased responsiveness of output to capital installed means a faster rate of growth for any given rate of increase in the quantity of capital.

In the third place, according to the productivity formula, capital-using/labour-saving non-neutral technological change leads to a movement along the new production function. The partial elasticity of output with respect to capital installed increases under such technological change. It is this which leads to an increase in the capital–labour ratio. Why does this capital–labour substitution occur?

The increased partial elasticity of output with respect to capital input means that, at the constant factor price ratio and at the initial capital–labour ratio, a net saving in cost can be made by substituting the now more 'productive' capital for labour. Furthermore, the decrease in the elasticity of output with respect to labour input (the decrease in a from a to a') means that the response of output to a reduction in labour input is not as great as previously. Hence it pays, in terms of reducing the unit cost of labour and capital combined, to substitute capital for labour and so the capital–labour ratio is increased. This 'mechanization' results in a further increase in labour productivity, the extent of this increase in labour productivity being measured by the ratio expressed in component (3).

This third component of capital-using/labour-saving non-neutral technological change is a very distinctive feature of this type of technological change. An increased capital-intensity of production is a diagnostic feature of capital-using/labour-saving technological change.

In the real world, whenever you see a technological change which leads to or results in a higher capital–labour ratio, or 'mechanization', then you must consider the possibility that the technological change is probably capital-using/labour-saving non-neutral.

(We have to be a little careful here because this is predicated on the basis of a constant factor price ratio. If there has been an increase in the price of labour relative to the price of capital, then this could also be a cause of increased capital-intensity of production. Dealing with the problem of an increased factor price ratio when non-neutral technological change occurs will be discussed at the end of the chapter.)

In the real world, the plant manager will simply be concerned with composite item (5) or the package of change as a whole. The situation measured by composite item (5) is simply that, in order to achieve production at minimum cost, the new technology is installed with a greater capital–labour ratio. What matters to the plant manager is the overall increase in labour productivity and the decrease in the unit cost of production. However, as economists we must look inside the package. We must understand composite item (5) analytically as the product of components (1), (2) and (3).

To illustrate all this we may again use the example of the production function for bricks.

Suppose that before a capital-using/labour-saving non-neutral technological change we had the following production function in labour productivity form where $Z_t = 212.1320344$ (see Chapter 2) and $b = 0.25$ (with the constant returns to scale restriction that $a + b = 1$):

$$(Q/L)_t = 212.1320344\ (K/L)_t^{0.25}$$

Now suppose that, after a capital-using/labour-saving non-neutral technological change, the adjustment factor Z is increased to 291.6815473 (the same 37.5 per cent increase as studied in the previous chapter in connection with neutral technological change so that we can use this example as a basis on which to build our understanding of non-neutral technological change), and suppose that the partial elasticity of output with respect to capital input, or the exponent on the capital input, is increased from $b = 0.25$ to $b' = 0.30$. Because of the restriction $a + b = 1$, this also means that the elasticity of output with respect to labour input, or the exponent on the labour input, is reduced from $a = 0.75$ to $a' = 0.70$. Accordingly, after the non-neutral technological change the labour productivity form of the production function is:

$$(Q/L)'_{t+1} = 291.6815473\ (K/L)'^{\,0.30}_{t+1}$$

If we have a price of labour p_L = \$2.40 per labour-hour and a price of capital p_K = 0.20 per annum, then the factor price ratio is $p_L/p_K = 12$. This factor price ratio is to be held constant, because this provides a basis from which to analyse the pure effect of technological change (that is, change in labour productivity *not* influenced by any mechanization stemming from an increasing relative expensiveness of labour).

The capital–labour ratio (at the minimum cost position) before technological change is $(K/L)^*_t = (p_L/p_K)/(a/b)$ = (\$2.40 per labour-hour/0.20 per annum)/(0.75/0.25) = \$4 per labour-hour per annum. The capital–labour ratio (at the minimum cost position) after capital-using/labour-saving non-neutral technological change is $(K/L)'^*_{t+1}$ = (\$2.40 per labour-hour/0.20 per annum)/(0.70/0.30) = \$5.142857143 per labour-hour per annum.

Consequently, before technological change, labour productivity at the minimum cost position (Plant I) is:

$$(Q/L)^*_t = 212.1320344 \times 4^{0.25}$$
$$= 300 \text{ bricks per labour-hour}$$

After capital-using/labour-saving non-neutral technological change, labour productivity at the minimum cost position (Plant III) is:

$$(Q/L)'^*_{t+1} = 291.6815473 \times 5.142857143^{0.30}$$
$$= 476.727597 \text{ bricks per labour-hour}$$

Thus this non-neutral technological change leads to an increase in labour productivity of 476.727597 bricks per labour-hour/300 bricks per labour-hour = 1.589, or 58.9 per cent.

Using the analytic formula for labour productivity change, we may break this proportionate increase in labour productivity down into three distinct components of non-neutral technological change as follows:

$$\frac{(Q/L)'^{*}_{t+1}}{(Q/L)^{*}_{t}} = \frac{Z_{t+1}}{Z_t} \times (K/L)^{*\,b'-b}_{t} \times \left(\frac{(K/L)'^{*}_{t+1}}{(K/L)^{*}_{t}}\right)^{b'}$$

$$= \frac{291.6815473}{212.1320344} \times 4^{0.30-0.25} \times \left(\frac{5.142857143}{4}\right)^{0.30}$$

$$= 1.37500 \times 1.07177 \times 1.07831$$

(1) Proportionate impact of neutral shift in production function

(2) Proportionate impact of exponent change on shift in production function

(3) Proportionate impact of non-neutral movement along new production function

(4) Proportionate impact of non-neutral shift in production function [(1) × (2)]

$$= 1.47368 \times 1.07831$$

(5) Proportionate impact of capital-using/labour-saving non-neutral technological change [(4) × (3)]

$$= 1.589$$

Thus, by applying the productivity formula, we can say that this particular capital-using/labour-saving non-neutral technological change shifted the labour productivity form of the production function by 47.368 per cent; that is, at the initial capital–labour ratio of $4 it raised labour productivity by 47.368 per cent. This change of 47.368 per cent can be seen as the product of two distinct components. First, the proportionate impact of the neutral shift in the production function due to the change in *Z*, or component (1), was 37.5 per cent. This is the same as the shift already studied in Chapter 2 in connection with neutral technological change. Second, the further proportionate impact of the exponent change on the shift in the production function, or component (2), was 7.177 per cent. (Note that $1.375 \times 1.07177 = 1.47368$.)

The non-neutral technological change also resulted in capital–labour substitution, thus increasing the minimum cost capital–labour ratio, and this 'mechanization' increased labour productivity by a further 7.831 per cent – component (3).

Accordingly, the total increase in labour productivity was 58.9 per cent. (Note that $1.47368 \times 1.07831 = 1.589$.)

Figure 3.1 illustrates the productivity formula. Figure 3.1 shows the labour productivity forms of the production function before and after the non-neutral change. For the sake of comparison, Figure 3.1 also shows the labour productivity form of the production function after a neutral technological change of 37.5 per cent (see Figure 2.3).

Before the technological change, Plant I is at point A operating with a capital–labour ratio of $4 per labour-hour per annum and labour productivity of 300 bricks per labour-hour.

After capital-using/labour-saving technological change, Plant III is at point E operating with an increased capital–labour ratio of $5.142857143 per labour-hour per annum and labour productivity of 476.727597 bricks per labour-hour.

The movement from A to E may be broken down into the three components of non-neutral technological change.

First, the shift from A to B illustrates the impact of the neutral shift in the production function – component (1). This illustrates what would happen if the adjustment factor increased by 37.5 per cent and nothing else happened – the case of neutral technological change considered in Chapter 2.

Second, the shift from B to D illustrates the impact of the exponent change on the shift in the production function. This is the additional and skew shift in the production function – component (2).

Third, the movement from D to E illustrates the impact of the non-neutral movement along the new production function – component (3). This results

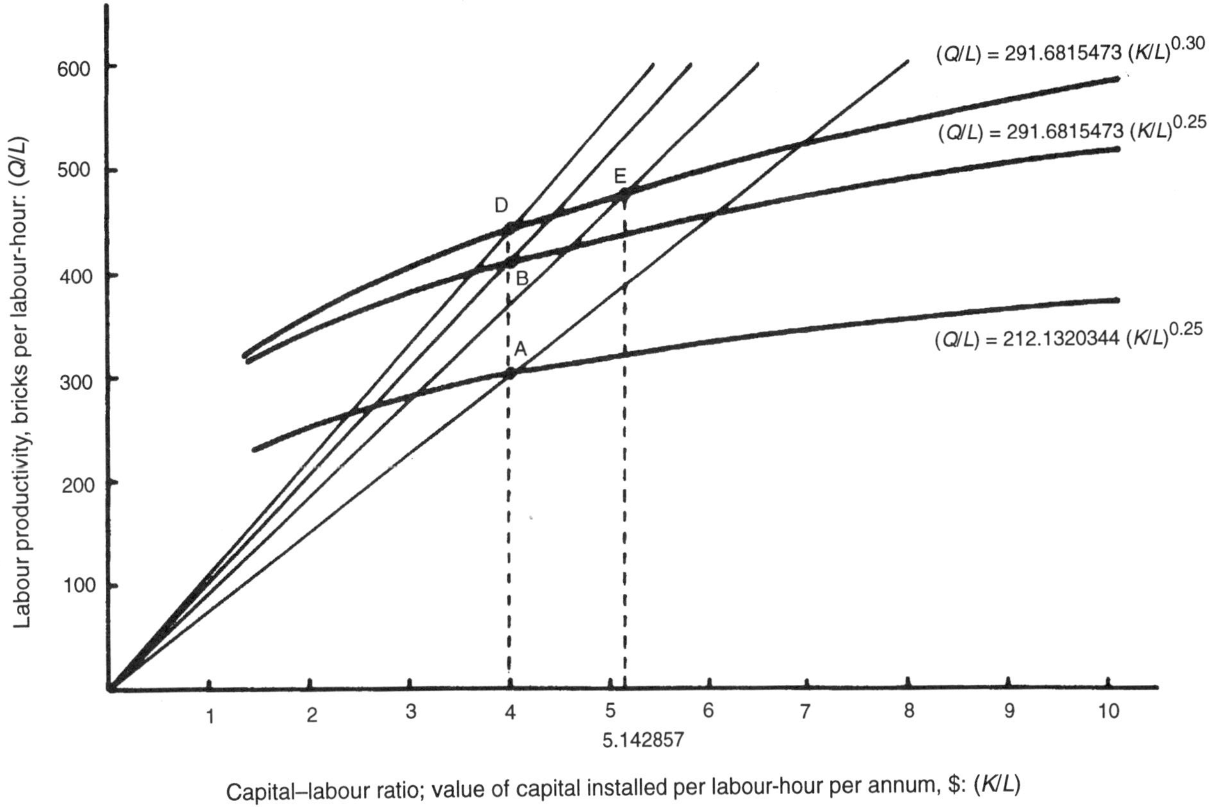

Figure 3.1 Neutral and non-neutral technological change

from substituting capital for labour in order to attain minimum cost production; unit labour and capital cost combined is a minimum at E (and is not a minimum at D).

Thus capital-using/labour-saving non-neutral technological change considerably enhances the increase in labour productivity. In this illustration, the non-neutral technological change, by comparison with what would occur under a neutral technological change, increases labour productivity from 412.5 bricks per labour-hour (after neutral technological change of 37.5 per cent – see Chapter 2) to 476.727597 bricks per labour-hour (after non-neutral technological change), or by 15.57 per cent.

Given the significance we place upon increases in labour productivity, because labour is the ultimately scarce factor of production and because labour productivity determines standards of living, it is important that we fully understand this 'labour productivity enhancing' aspect of capital-using/labour-saving non-neutral technological change.

However, what matters to the plant manager or production engineer is the overall increase in labour productivity of 58.9 per cent and the consequent reduction in unit cost.

3.3 Unit cost: before and after comparison

Non-neutral technological change is capable of having a considerable impact on the unit cost of labour and capital combined (at the minimum cost position), denoted v^*. We have, in general (without the time or 'vintage' subscript):

$$v^* = p_L (L/Q)^* + p_K (K/Q)^*$$

Unit labour requirement, $(L/Q)^*$, may be calculated as the inverse of labour productivity $(Q/L)^*$. Unit capital requirement, $(K/Q)^*$ may be calculated as the capital–labour ratio multiplied by unit labour requirement; that is: $(K/Q)^* = (K/L)^*(L/Q)^*$. For the non-neutral technological change just considered, we have (units of measurement taken as read, and for p_L = \$2.40 per labour-hour and $p_K = 0.20$ per annum):

	Before technological change, t	*After capital-using/ labour-saving non-neutral technological change, t + 1*
$(Q/L)^*$	300	476.727597
$(L/Q)^*$	0.0033333	0.00209763

$(K/L)^*$	4	5.14285714
$(K/Q)^*$	0.0133333	0.01078783
v^*	0.0106667	0.00719189

Consequently, non-neutral technological change has reduced unit labour and capital cost combined in the ratio 0.6742 (= 0.00719189/0.0106667), or by 32.6 per cent.

We now show that the unit cost of labour and capital combined, v^*, and thence also the proportionate reduction in v^*, may be calculated from the parameters of the production function together with the factor price ratio. This is very useful because it means that in the absence of information on the quantity of output and the quantities of factor inputs we can nevertheless calculate the proportionate impact on unit labour and capital cost combined of any non-neutral technological change.

In general, from the labour productivity form of the production function $(Q/L)^* = Z\,(K/L)^{*b}$ we may derive:

$$(L/Q)^* \quad = \quad Z^{-1}\,(K/L)^{*-b}$$

But $(K/L)^* = (p_L/p_K)/(a/b)$ and correspondingly for $(K/L)'^*$ with (a'/b'). Therefore, by substitution:

$$(L/Q)^* \quad = \quad Z^{-1}\,[(p_L/p_K)/(a/b)]^{-b}$$

Unit capital requirement may be calculated as follows:

$$\begin{aligned}(K/Q)^* \quad &= \quad (K/L)^*(L/Q)^* \\ &= \{(p_L/p_K)/(a/b)\}\,\{Z^{-1}[(p_L/p_K)/(a/b)]^{-b}\} \\ &= Z^{-1}\,[(p_L/p_K)/(a/b)]^{1-b}\end{aligned}$$

Unit labour and capital cost combined at the minimum cost position, v^*, may be calculated as follows:

$$\begin{aligned}v^* \quad &= \quad p_L\,(L/Q)^* + p_K\,(K/Q)^* \\ &= \quad p_L\,Z^{-1}\,[(p_L/p_K)/(a/b)]^{-b} + p_K\,Z^{-1}\,[(p_L/p_K)/(a/b)]^{1-b} \\ &= \quad Z^{-1}\,p_K\,\{(p_L/p_K)[(p_L/p_K)/(a/b)]^{-b} + [(p_L/p_K)/(a/b)]^{1-b}\} \\ &= \quad Z^{-1}\,p_K\,\{(p_L/p_K)^{1-b}(a/b)^{b} + (p_L/p_K)^{1-b}(a/b)^{b-1}\} \\ &= \quad Z^{-1}\,p_K\,(p_L/p_K)^{1-b}\,\{(a/b)^{b} + (a/b)^{b-1}\}\end{aligned}$$

We have the following values: $Z_t = 212.1320344$; $Z_{t+1} = 291.6815473$; $p_L =$ \$2.40 per labour-hour; $p_K = 0.20$ per annum; $p_L/p_K =$ \$2.40/0.20 = 12; $a = 0.75$; $b = 0.25$; $a/b = 0.75/0.25 = 3$; $a' = 0.70$; $b' = 0.30$; and $a'/b' = 0.70/0.30 = 2.3333333$. Using these values:

$$\begin{aligned} v^*_t &= 212.1320344^{-1} \times 0.20 \times 12^{1-0.25} \{3^{0.25} + 3^{0.25-1}\} \\ &= \$0.0106667 \text{ per brick} \\ &= \$10.6667 \text{ per thousand bricks} \end{aligned}$$

and:

$$\begin{aligned} v'^*_{t+1} &= 291.6815473^{-1} \times 0.20 \times 12^{1-0.30}\{2.3333333^{0.30} + 2.3333333^{0.30-1}\} \\ &= \$0.00719189 \text{ per brick} \\ &= \$7.19189 \text{ per thousand bricks} \end{aligned}$$

This is a substantial reduction in unit labour and capital cost combined.

The proportionate reduction in unit labour and capital cost combined can also be computed from a formula involving only the production function parameters and the factor price ratio. We proceed as follows.

Insert time subscripts and the prime subscript, and substitute $1 - b$ for a and $1 - b'$ for a' (because in the labour productivity form of the production function we use only the exponent b), then the proportionate reduction in unit labour and capital cost combined is:

$$\frac{v'^*_{t+1}}{v^*_t} = \frac{Z_{t+1}^{-1}\, p_K (p_L/p_K)^{1-b'} \{(a'/b')^{b'} + (a'/b')^{b'-1}\}}{Z_t^{-1}\, p_K (p_L/p_K)^{1-b} \{(a/b)^{b} + (a/b)^{b-1}\}}$$

$$= \frac{Z_t}{Z_{t+1}} \times (p_L/p_K)^{b-b'} \times \frac{\left(\frac{1-b'}{b'}\right)^{b'} + \left(\frac{1-b'}{b'}\right)^{b'-1}}{\left(\frac{1-b}{b}\right)^{b} + \left(\frac{1-b}{b}\right)^{b-1}}$$

This formula (that is, everything on the right-hand side of the equals sign) I call the *analytic formula for the difference in unit labour and capital cost combined*, or the 'v'^*/v^* formula' for short.

Using the same values for the parameters, we may apply the v'^*/v^* formula to get:

$$\frac{v'^*_{t+1}}{v^*_t} =$$

$$\frac{212.1320344}{291.6815473} \times 12^{0.25-0.30} \times \frac{\left(\frac{1-0.30}{0.30}\right)^{0.30} + \left(\frac{1-0.30}{0.30}\right)^{0.30-1}}{\left(\frac{1-0.25}{0.25}\right)^{0.25} + \left(\frac{1-0.25}{0.25}\right)^{0.25-1}}$$

$$= 0.727272 \times 0.883163 \times 1.049726$$

$$= 0.6742$$

Thus the proportionate reduction (ratio form) in unit cost of labour and capital combined (at the minimum cost position) can be calculated from the production function parameters and the factor price ratio.

The reduction in unit labour and capital cost combined is substantial. Taking a unit materials cost of \$9.33 per thousand bricks, the unit total cost would change from \$20 per thousand bricks (= \$10.67 per thousand bricks + \$9.33 per thousand bricks) to \$16.52 per thousand bricks (= \$7.19 per thousand bricks + \$9.33 per thousand bricks), and taking an assumed constant product price p_Q = \$22 per thousand bricks, the unit net margin would increase from \$2.00 per thousand bricks before technological change (= \$22 per thousand bricks – \$20 per thousand bricks) to \$5.48 per thousand bricks after capital-using/labour-saving non-neutral technological change (= \$22 per thousand bricks – \$16.52 per thousand bricks). The unit capital requirement being reduced from \$13.3333 per thousand bricks per annum to \$10.78783 per thousand bricks per annum, the efficiency rate of profit would increase from 15 per cent per annum to 50.8 per cent per annum.

This indicates the potential of non-neutral technological change for increasing profitability, although in the real world a reduction in product price would share some, and perhaps a great part, of the gains with purchasers of the product which has been the subject of the process innovation.

3.4 Neutral/non-neutral comparison

For a full understanding of non-neutral technological change we have to consider whether non-neutral technological change may offer a systematic advantage over neutral technological change. If non-neutral technological change were systematically more advantageous than neutral technological change, then it is likely that the direction of invention and innovation would be deliberately non-neutral, so enhancing labour productivity growth both historically and prospectively.

In terms of Figure 3.1, we have to analyse the difference between point B after a neutral technological change and point E after a non-neutral technological change with the same neutral shift in the production function.

It is clear that both neutral and non-neutral technological change result in an increase in multifactor productivity measured by the increase in the adjustment factor Z or the proportionate impact of the neutral shift in the production function (component (1) of the analytic formula for labour productivity change).

The increase in Z simultaneously reduces *both* unit labour requirement *and* unit capital requirement, thus reducing unit labour and capital cost combined. The proportionate effect of this on unit labour and capital cost combined is shown in the v'^*/v^* formula by the ratio Z_t/Z_{t+1}.

The unit capital requirement may be represented *inversely* in Figure 3.1 by the slope of a ray from the origin to the labour productivity form of the production function. Any such ray has the slope $(Q/L)/(K/L) = (Q/K)$, and (Q/K) or capital productivity is the inverse of the unit capital requirement, (K/Q).

Accordingly, the more steep is the slope of any such ray, the *smaller* is the unit capital requirement and, conversely, the less steep is the slope of any such ray, the *greater* is the unit capital requirement.

In Figure 3.1 we can see that in the movement from A to B, the neutral shift in the production function, labour productivity increases (unit labour requirement decreases) in the upward vertical movement (measured at the initial capital–labour ratio) and *at the same time* capital productivity increases (unit capital requirement decreases) in the increased slope of the ray from the origin (through A and B respectively).

Thus, as we already know, both neutral and non-neutral technological change result in an advantage, over the situation before technological change, in unit labour and capital cost combined. But the important question is whether or not non-neutral technological change (capital-using/labour-saving) results, *systematically*, in any additional or further cost saving.

To answer this question we must abstract from the change in Z and concentrate upon the remaining proportionate change in unit labour and capital cost combined. (This is why we assume the same value of Z_{t+1} for both neutral and non-neutral technological change: the important question for the neutral/non-neutral comparison is what happens in addition to the change in Z.)

In terms of Figure 3.1, the remaining proportionate difference in unit labour and capital cost combined must be due to the combination of the additional skew shift in the production function (the movement from B to D) or the impact of the exponent change on the shift in the production function (component (2) of the analytic formula for labour productivity change) and the non-neutral movement along the new production function (the movement from D to E) or 'mechanization' due to the exponent change (component (3) of the analytic formula for labour productivity change).

In terms of the v'^*/v^* formula, the remaining proportionate difference in unit labour and capital cost combined is measured by the product of the two expressions other than Z_t/Z_{t+1}, but the following explanation will proceed by using Figure 3.1. However, the v'^*/v^* formula does reveal that one of the conditions required for the following analysis is that the factor price ratio be a number greater than 1. In economic terms, a greater importance must attach to saving a unit of labour than to using a unit of capital.

It is clear that labour productivity will be higher at D than at B, and hence unit labour requirement and unit labour cost will be lower at D than at B.

It is also clear that capital productivity will be higher at D than at B (the slope of the ray through D is steeper than the slope of the ray through B), and hence unit capital requirement and unit capital cost will be lower at D than at B.

Together this means that unit labour and capital cost combined *must* be lower at D than at B.

But position D is not where the process of non-neutral technological change ends because, owing to the change in (the ratio of) the exponents, D is not a position of minimum cost production on the new production function. On the new production function, minimum unit labour and capital cost combined occurs at point E, corresponding to $(K/L)'^*_{t+1}$.

Consequently, under capital-using/labour-saving non-neutral technological change there will be a non-neutral movement along the new production function from D to E (from $(K/L)^*_t$ to $(K/L)'^*_{t+1}$).

It is clear that labour productivity will be higher at E than at D, and hence unit labour requirement and unit labour cost will be lower at E than at D.

But we can see that capital productivity will be lower at E than at D and hence unit capital requirement and unit capital cost will be *higher* at E than at D.

However, because D is not a position of minimum unit cost of labour and capital combined, and because E is the position of minimum unit cost of labour and capital combined, the (absolute value of the) reduction in unit labour cost when moving from D to E must be greater than the increase in unit capital cost when moving from D to E.

Consequently, unit labour and capital cost combined must be lower at E than at D.

If unit labour and capital cost combined must be lower at D than at B, and if unit labour and capital cost combined must be lower at E than at D, then it follows that unit labour and capital cost must be lower at E than at B.

All this may be illustrated by the data in Table 3.1, giving unit labour requirement, unit capital requirement, unit labour cost, unit capital cost, unit labour and capital cost combined, and the unit net margin for each of positions B, D, and E according to the illustrative values used in this chapter (including an assumed constant unit materials cost of $9.33 per thousand bricks and an assumed constant product price of $22 per thousand bricks).

Positions B and E are positions of minimum unit labour and capital cost combined on their respective production functions; position D is not a position of minimum unit labour and capital cost combined on its production function.

We can see the following features in Table 3.1.

Unit labour requirement, and hence unit labour cost, is lower at D than at B. Furthermore, unit capital requirement, and hence unit capital cost, is lower at D than at B.

Therefore, unit labour and capital cost combined is (and must be) lower at D than at B.

Unit labour requirement, and hence unit labour cost, is lower at E than at D.

But unit capital requirement and hence unit capital cost is higher at E than at D. However, D is not a position of minimum unit labour and capital cost combined on the new production function. Accordingly, in the movement from D to E, the reduction in unit labour cost, –$0.395 per thousand bricks, is greater (in absolute value) than the increase in unit capital cost, $0.348 per thousand bricks.

Consequently, unit labour and capital cost combined is (and must be) lower at E than at D.

Table 3.1
The neutral/non-neutral comparison (hypothetical data)

	Per 1,000 bricks[a]			Change in unit cost or unit net margin, $ per 1,000 bricks		
	After neutral technological change, Position B	After non-neutral technological change				
		Position D	Position E	From B to D	From D to E	From B to E
Unit labour requirement, labour-hours per 1,000 bricks, (L/Q) 1,000[b]	2.424	2.262	2.098	–	–	–
Unit capital requirement, $ per 1,000 bricks p.a., (K/Q) 1,000[c]	9.697	9.048	10.788	–	–	–
Unit labour cost, $ per 1,000 bricks, $2.40 (L/Q) 1,000	5.818	5.429	5.034	–0.389	–0.395	–0.784
Unit capital cost, $ per 1,000 bricks, 0.20 (K/Q) 1,000	1.939	1.810	2.158	–0.129	+0.348	+0.219
Unit labour and capital cost combined, $ per 1,000 bricks, v	7.757	7.239	7.192	–0.518	–0.047	–0.565
Unit net margin, $ per 1,000 bricks[d]	4.913	5.431	5.478	+0.518	+0.047	+0.565

(*a*) Calculations up to and including v use unrounded figures.
(*b*) Computed from the labour productivity production function, with relevant value of (K/L).
(*c*) Computed as the relevant capital–labour ratio multiplied by unit labour requirement.
(*d*) At an assumed constant unit materials cost of $9.33 per thousand bricks and an assumed constant product price of $22 per thousand bricks.

Source: calculated from formulae and parameter values given in text.

Consequently also, unit labour and capital cost combined is (and must be) lower at E than at B.

In the last column of Table 3.1, we can see that in the overall movement from B to E, or in the overall neutral/non-neutral comparison, unit labour cost has decreased (by –$0.784 per thousand bricks), while unit capital cost has increased (by $0.219 per thousand bricks).

Nevertheless, the balance between these two changes must be that unit labour and capital cost combined decreases overall. This we can see in the second-last row of Table 3.1, because the change in unit labour and capital cost combined between B and D must be negative, and the change in unit labour and capital cost combined between D and E must also be negative.

Capital-using/labour-saving non-neutral technological change provides a systematic cost advantage over neutral technological change (given that both changes are subject to the same neutral shift in the production function). This analysis is predicated upon the particular form of the production function we have used and upon the factor price ratio being greater than 1, but we should not be surprised to find that the history (past and prospective) of technological change is dominated by capital-using/labour-saving non-neutral technological change.

While the neutral/non-neutral comparison leads to a definite conclusion about the favourable advantage of unit labour and capital cost combined, the same is not the case for the efficiency rate of profit.

At a constant product price, the unit net margin will definitely be greater under capital-using/labour-saving non-neutral technological change than under neutral technological change.

However, the unit capital requirement is likely also to be greater under such non-neutral technological change (there is limited scope for the unit capital requirement at the minimum cost position to be lower, but this is a less likely outcome).

What happens to the efficiency rate of profit depends on the balance between the proportionate increase in the unit net margin (which increase is certain to occur if the product price is constant) and the proportionate increase in the unit capital requirement (which increase is likely to occur). This balance may go either way; in the illustrative case considered, at the assumed unit materials cost of $9.33 per thousand bricks and the assumed product price of $22 per thousand bricks, the unit net margin is 11.5 per cent greater under non-neutral technological change than under neutral technological change, but the unit capital requirement is 11.3 per cent greater, so the efficiency rate of profit is much the same under this non-neutral technological change (50.8

per cent per annum (= \$5.478 per thousand bricks/\$10.788 per thousand bricks per annum)) than it would be under a comparable neutral technological change (50.7 per cent per annum (= \$4.913 per thousand bricks/\$9.697 per thousand bricks per annum)). But the cost advantage is the important feature of the neutral/non-neutral comparison.

3.5 Non-neutral technological change and the factor price ratio

In making a before and after comparison involving non-neutral technological change we may find that, over the relevant period of time, the factor price ratio has changed, thus leading to mechanization additional to that which would be expected from non-neutral technological change in the absence of a change in the factor price ratio. How may we deal with the effect of this additional mechanization when analysing non-neutral technological change?

For analysing the difference in labour productivity made by the non-neutral technological change, we previously derived an analytic formula for the labour productivity difference. This formula included the actual minimum cost capital–labour ratio after the non-neutral technological change, $(K/L)'^*_{t+1}$. If there is a change in the factor price ratio, $(K/L)'^*_{t+1}$ will be determined not only by the change in the production function's exponents but also by the change in the factor price ratio. Denote the changed factor price ratio by the subscript $t + 1$, then:

$$(K/L)'^*_{t+1} = (p_L/p_K)_{t+1}/(a'/b')$$

Accordingly, the productivity formula as previously presented would incorporate the impact on labour productivity of the additional mechanization due to the changed factor price ratio.

If we wish to exclude the impact of the additional mechanization, we may do this by calculating an hypothetical minimum cost capital–labour ratio as this capital–labour ratio would have been with the (given) change in the exponents to the production function but at the initial factor price ratio, denoted $(p_L/p_K)_t$. Let us denote this hypothetical capital–labour ratio as $(K/L)''^*_{t+1}$ where:

$$(K/L)''^*_{t+1} = (p_L/p_K)_t/(a'/b')$$

This hypothetical capital–labour ratio will result in an hypothetical level of labour productivity due solely to the non-neutral technological change; this hypothetical labour productivity may be denoted $(Q/L)''^*_{t+1}$.

The analytic formula for labour productivity change, excluding the impact of the additional mechanization due to the change in the factor price ratio, may now be derived by a similar route as:

$$\frac{(Q/L)''\,*_{t+1}}{(Q/L)\,*_t} = \frac{Z_{t+1}}{Z_t} \times (K/L)*_t^{\,b'-b} \times \left(\frac{(K/L)''\,*_{t+1}}{(K/L)\,*_t}\right)^{b'}$$

In the next chapter we shall use this formula for estimating the impact on labour productivity of non-neutral technological change in the United States railroad industry.

Suppose we now wish to estimate the overall reduction in unit labour and capital cost combined when there is not only a non-neutral technological change but also a change in the factor price ratio. How may this be done?

The answer is quite straightforward. If you inspect the analytic formula for the difference in unit labour and capital cost combined, you will see that we were able to simplify the term in (p_L/p_K) in both numerator and denominator.

However, if the factor price ratio changes because of a change in p_L only with p_K unchanged, then the $v'*/v*$ formula cannot have this particular simplification. Denoting the factor price ratio before technological change as $(p_L/p_K)_t$ and the changed factor price ratio after the non-neutral technological change as $(p_L/p_K)_{t+1}$, we have by the same derivation:

$$\frac{v'*_{t+1}}{v*_t} = \frac{Z_t}{Z_{t+1}} \times \frac{(p_L/p_K)_{t+1}^{\,1-b'}}{(p_L/p_K)_t^{\,1-b}} \times \frac{\left(\frac{1-b'}{b'}\right)^{b'} + \left(\frac{1-b'}{b'}\right)^{b'-1}}{\left(\frac{1-b}{b}\right)^{b} + \left(\frac{1-b}{b}\right)^{b-1}}$$

This formula then gives the cost decrease (in ratio form) for unit labour and capital cost combined resulting both from non-neutral technological change and also from the further mechanization in response to the changed factor price ratio.

In the next chapter, we use this formula to compute the change in unit labour and capital cost combined in the United States railroad industry over the period 1948 to 1967.

4 Non-neutral Technological Change in the United States Railroad Industry

4.1 Diesel–electric locomotives replace steam locomotives

Romantic and glamorous though they may seem in nostalgic retrospect, steam locomotives were very inefficient in all sorts of ways. The external combustion engine wastes a great deal of heat by comparison with the internal combustion engine. Steam locomotives required a considerable amount of labour to start (to fire up) and then to keep fuelled and watered. Steam locomotives also required much labour-time to clean and to maintain.

From its beginnings in the 1890s the diesel engine was designed by Rudolf Diesel on thermodynamic principles to be a fuel-efficient internal combustion engine. Diesel's use of a high compression ratio was deliberately intended to provide the highest possible thermal efficiency and led also to compression-ignition, which is a distinctive feature of this engine.

The mid-twentieth century adaptation of the modern high-speed diesel engine to providing tractive power in the railroad industry was a major technological change which economized on both fuel and labour.

A diesel locomotive should more properly be called a diesel–electric locomotive, because in a railroad locomotive the diesel engine drives a generator which supplies electric power to several (large) electric motors, each of which drives a single unit of one axle with two wheels.

This arrangement is, for technical reasons, superior to a direct mechanical linkage between the diesel engine and the axles: electric motors can be attached to many more than one or two axles, thus providing more driving power; electric motors can cope with the temporary overload of force required to get a train moving from a stationary position while a mechanical driveshaft of reasonable weight cannot cope with the strain of getting the train moving. Essentially, the electrical transmission of power protects the diesel engine itself from shock loads whereas with mechanical transmission the shocks are transmitted to the engine. Additionally, the gearbox which would be required for mechanical linkage would be too complex and heavy (especially when more than one axle is involved); and a mechanical linkage involves a greater loss of power in transmission.

However, we shall follow common usage in referring to a diesel–electric locomotive simply as a 'diesel locomotive'.

The replacement of steam locomotives by diesel locomotives was a phased process which took place in the railroad industry mainly in the 1940s and 1950s.

During the 1960s, 'first-generation' diesel locomotives were themselves replaced by technologically improved and much more powerful 'second-generation' diesel locomotives.

First-generation diesel locomotives were themselves more powerful than steam locomotives, and were much more efficient in the use of fuel: a diesel locomotive 'delivers' about 35–40 per cent of the fuel it consumes as pulling power on the track, while a steam locomotive 'delivered' only about 5 per cent. In other words, a diesel locomotive was at least seven times more fuel efficient than a steam locomotive. This is very significant in the railroad industry, where fuel is one of the most important inputs.

Furthermore, after any long trip, a diesel locomotive could be refuelled and working again in half an hour or less, while a steam locomotive required at least two hours of elapsed time, and often more, to clean out its fire-box and take on coal and water.

A diesel locomotive also uses fuel (petroleum distillate) with a higher heat content per pound avoirdupois than coal, and this permits a diesel locomotive to carry fuel for a much longer trip.

In 1961, second-generation diesel locomotives began to be introduced in the United States railroad industry. These were technologically more advanced, required less maintenance, and were more powerful, ranging between 2,500 and 6,000 horsepower by comparison with 1,200 to 1,500 horsepower for the first-generation diesel locomotives.

In 1963, six-axle diesel locomotives (an electric motor attached to each of six axles) began to be introduced. These locomotives were better able to utilize the greater horsepower of the diesel engine and this further increased the pulling power of the locomotive.

In 1956, the acquisition cost of a diesel locomotive was about three-and-three-quarters as much as the acquisition cost of a steam locomotive. But a diesel locomotive could do more than twice – or even treble or quadruple –

the work of a steam locomotive, and this alone largely offset the extra acquisition cost.

Furthermore, diesel locomotives offered many more cost savings than simply on fuel.

Diesel locomotives saved considerably on labour (eliminating the labour required to start the locomotive, to keep it fuelled and watered, and to clean it). Diesel locomotives also required fewer labour-hours for maintenance.

In 1956, it was estimated that the unit cost per ton-mile of diesel locomotive haulage was 71 per cent less (!) than the unit cost of steam locomotive haulage.

We can study the progress of this technological change in a valuable report by the United States Department of Labor Bureau of Labor Statistics, *Railroad Technology and Manpower in the 1970's*, Bulletin 1717, 1972 (hereafter cited as Bulletin 1717).

The progress of dieselization in the United States inter-city railroad industry may be shown in the following statistics from Bulletin 1717 (pp. 14, 71 and 76):

			Index, 1948 = 100	
	Total number of locomotives, thousands	*Percentage of total diesel*	*Quantity of annual output*[(a)]	*Locomotive productivity*
1948	41.9	19	100	100
1957	30.2	90	93	129
1962	28.6	98	88	129
1967	27.7	99	104	157

(*a*) Index of weighted revenue passenger-miles (passenger traffic) and weighted revenue ton-miles (freight traffic).

The change in locomotive productivity between 1948 and 1957 is due largely to the introduction of first-generation diesel locomotives; the change in locomotive productivity between 1962 and 1967 is due largely to the introduction of second-generation diesel locomotives.

This 'new-plant' technological change was supplemented by many other (largely 'in-plant') technological changes, such as containerized freight, improved traffic control, and improved 'maintenance of way' (repairs to and maintenance of track), but the major changes were the replacement of steam locomotives by diesel locomotives, the replacement of first-generation diesel locomotives by second-generation diesel locomotives, and the related changes (longer trains capable of pulling more load at greater average speed) which this dieselization made possible.

4.2 Labour productivity in the railroad industry

How did all this affect labour productivity?

Tables 4.1 and 4.2 provide the basic data to show, first, the impact of technological change in the railroad industry on labour productivity, and, second, that this technological change was capital-using/labour-saving non-neutral technological change.

For the selected years 1948, 1952, 1957, 1962 and 1967 (the only years for which data on the value of capital installed are available), the number of railroad employees declined consistently.

Over the period as a whole, the decrease in employment, from 1.517 million employees in 1948 to 0.698 million employees in 1967, was 54 per cent (or about 4 per cent per annum compound).

The quantity of labour input should, preferably, be measured in labour-hours per annum. The data show that the quantity of labour-hours per annum declined from 3.5 billion in 1948 to 1.5 billion in 1967, a decline of 57 per cent (or about 4.3 per cent per annum compound).

The annual quantity of output in the railroad industry, measured by value added at constant prices, at first declined and then increased to be about 7 per cent greater in 1967 than in 1948. Allowing for this fluctuation, total annual output did not change by much.

The value of capital installed in the railroad industry (at constant prices) increased slightly. Although a diesel locomotive had a greater acquisition cost than a steam locomotive, far fewer of them were required to do the same work. For example, it was recorded that on the Australian trans-continental rail line in 1956, eleven diesel locomotives were required to do the work formerly requiring forty-three steam locomotives!

The impact of these changes on factor productivity and the capital-intensity of production is shown in Table 4.2.

Unit labour requirement over the period as a whole decreased by 60 per cent, or 4.7 per cent per annum compound. Conversely, labour productivity increased by 4.9 per cent per annum compound.

Unit capital requirement increased slightly with the introduction of first-generation diesel locomotives but then declined slightly with the introduction of second-generation diesel locomotives.

Table 4.1
Factor inputs and output in the United States railroad industry, selected years 1948 to 1967

Year	All railroad employees,[a] million	Average weekly hours, Class I railroads	Estimated labour input,[b] million labour-hours per annum, *L* /1,000,000	$ million 1958 prices: Value added,[c] *V* /1,000,000	$ million 1958 prices: Value of railroad and equipment,[d] *K* /1,000,000
1948	1.517	46.2	3,504.270	10,500	51,900
1952	1.400	40.6	2,842.000	10,100	53,600
1957	1.121	41.7	2,337.285	9,500	53,300
1962	0.796	42.6	1,695.480	9,200	51,000
1967	0.698	43.2	1,507.680	11,200	52,100

(*a*) Apparently in March of each year; see Bulletin 1717, Appendix IV, p. 83.

(*b*) Based on an assumed 2 weeks of vacation/holidays per annum, or 50 weeks' work during the year. In the absence of a time series on hours worked per annum, this figure appears empirically to be an appropriate average for the period as a whole. For union nonsupervisory employees, the initial provision in 1942 of vacation pay was for '1 week's vacation after 1 year of service, but additional weeks were added in 1945, 1954, and 1965'. 'The vacation policy at the end of 1967 for union nonsupervisory employees was 1 year of service – 1 week of vacation; 3 years – 2 weeks; 10 years – 3 weeks; and 20 years – 4 weeks.' 'Pay and Fringe Benefits of Railroad Employees', *Monthly Labor Review,* September 1968, Vol. 91, No. 9, pp. 46–7. Accordingly, labour-hours per annum is calculated as: All railroad employees *multiplied by* Average weekly hours *multiplied by* 50 weeks' work per annum per employee. This assumes that the average weekly hours worked by all railroad employees was equal, or at least very similar, to that worked by Class I railroad employees.

(*c*) For all railroads; see Bulletin 1717, table 8, p. 80, note 2.

(*d*) Apparently on January 1 of year following (see Bulletin 1717, p. 79). This apparently relates to data on the number of railroad employees, because Bulletin 1717 (p. 37) divides the former by the latter to give 'the value of road [track] and equipment per employee'.

Sources: United States Department of Labor Bureau of Labor Statistics, *Railroad Technology and Manpower in the 1970's*, Bulletin 1717, 1972, tables 2, 18, Appendix III, table 8, Appendix IV, table 1, pp. 7, 37, 80, 83; average weekly hours: United States Department of Commerce Bureau of the Census, *Historical Statistics of the United States Colonial Times to 1970*, Bicentennial Edition, Part 1, series D 816, p. 170 (see also: United States Department of Labor Bureau of Labor Statistics, *Employment and Earnings Statistics for the United States 1909–65*, Bulletin No. 1312–3, December 1965, p. 576).

Table 4.2
Impact of non-neutral technological change in the United States railroad industry, 1948 to 1967

Year	Unit labour requirement, labour-hours per dollar of 1958 price value added, *L/V*	Unit capital requirement, 1958 dollars of capital installed per dollar of 1958 price value added per annum, *K/V*	$, 1958 prices	
			Labour productivity, $ of 1958 price value added per labour-hour, *V/L*	Capital–labour ratio, 1958 dollars of capital installed per labour-hour per annum, *K/L*
1948	0.333740	4.942857	2.996344	14.810503
1952	0.281386	5.306931	3.553835	18.859958
1957	0.246030	5.610526	4.064545	22.804237
1962	0.184291	5.543478	5.426192	30.079977
1967	0.134614	4.651786	7.428632	34.556405

Source: calculated from Table 4.1.

A big change occurred in the capital–labour ratio which, at constant 1958 prices, increased from $14.81 of capital installed per labour-hour per annum in 1948 to $34.56 per labour-hour per annum in 1967. This represents an increase in the capital–labour ratio over the period as a whole of 133 per cent, or 4.6 per cent per annum compound.

At first sight, this is evidence that technological change in the railroad industry was capital-using/labour-saving non-neutral technological change, and we can use production function analysis to show that this was indeed the case.

4.3 Dieselization as capital-using/labour-saving non-neutral technological change

We may describe and analyse the nature of technological change in the railroad industry through the production function, showing the changes over the years in the unit isoquants and in the labour productivity form of the production function.

In order to do this we must estimate values for the parameters of the production function, namely: a, b, and Z' (using Z' to signify the adjustment factor because *value added at constant prices* is used to measure the 'quantity' of output). If we assume a constant returns to scale production function, then an estimated value of b implies an estimated value of a because $1 - b = a$. We shall assume that, by and large, constant returns to scale more or less prevailed in the railroad industry, although there is some evidence that in certain parts of its operations, such as 'marshalling' or 'classification' yards, economies of scale did prevail and that advantage was taken of such economies of scale by consolidation of facilities. However, for illustrative purposes the assumption of constant returns to scale may be accepted as reasonably applicable to the railroad industry.

We know that the capital–labour ratio at the minimum cost position is determined by the value of the factor price ratio divided by the ratio of the exponents to the production function. That is, $(K/L)^* = (p_L/p_K)/(a/b)$. From this, and under constant returns to scale so that $a = 1 - b$, we may by rearrangement derive that:

$$b = \frac{1}{1 + (p_L/p_K)/((K/L)^*)}$$

In Table 4.2 we have values of (K/L) for each of the given years. We shall assume that these values are equal to, or are reasonably close to, minimum cost capital–labour ratios, and we accordingly take each of these values as estimates of $(K/L)^*$ in the railroad industry.

Table 4.3 gives, for the selected years, data on average hourly earnings in the railroad industry, adjusted for price inflation by using the gross national product implicit deflator for 1958 = 100 (this brings the average hourly earnings, or the price of labour, into conformity with the constant price data for value added and the capital stock). These 'real' average hourly earnings are used as the estimated values of p_L.

Bulletin 1717 (p. 79) gives an average 40-year life for railroad track and equipment. We can thus derive an estimated value of $p_K = 1/40$ years $= 0.025$ per annum. We assume that this value of the annual rate of depreciation, or p_K, holds for each year.

We can thus estimate, in the fourth column of Table 4.3, values of the factor price ratio, p_L/p_K, for each of the given years. The factor price ratio increases over the period as a whole, and so we should expect some additional capital–labour substitution or mechanization in response to this increase in the relative price of labour.

Table 4.3
The capital-using/labour-saving non-neutral nature of technological change in the United States railroad industry, 1948 to 1967

Year, t	Price index, gross national product implicit price deflator, 1958 = 100	Average hourly earnings, Class I railroad employees,[a] Current prices, $	Average hourly earnings, Class I railroad employees,[a] $ at 1958 prices, $(p_L)_t$	Factor price ratio,[b] $(p_L/p_K)_t$	Estimate of partial elasticity of output with respect to capital input,[c] b_t	Estimate of adjustment factor,[d] Z'_t
1948	79.6	1.30	1.633166	65.326633	0.184814	1.820763
1952	87.5	1.83	2.091429	83.657143	0.183969	2.070320
1957	97.5	2.26	2.317949	92.717949	0.197401	2.192491
1962	105.8	2.72	2.570888	102.835539	0.226309	2.511584
1967	117.6	3.24	2.755102	110.204082	0.238714	3.188893

(*a*) All employees except executives, officials, and staff assistants. Following procedure assumes average hourly earnings for all railroad employees is equal to, or at least close to, average hourly earnings for Class I railroad nonsupervisory employees.

(*b*) Using a constant value for $p_K = 0.025 = 1/40$. The figure of a 40-year life for capital is given in Bulletin 1717, p. 79; calculation uses unrounded values of p_L at 1958 prices.

(*c*) The estimate for b_t is calculated as: $b_t = \dfrac{1}{1 + ((p_L/p_K)_t/(K/L)^*_t)}$ using values of $(K/L)_t$ from Table 4.2 as calculated from Table 4.1; calculation uses unrounded values of $(p_L/p_K)_t$ and $(K/L)_t$.

(*d*) The estimate for Z'_t is calculated as: $Z'_t = \dfrac{(V/L)_t}{(K/L)_t^{b(t)}}$ using values of $(V/L)_t$ from Table 4.2 and noting that the estimate of b_t appropriate to each year t is used. Calculation uses unrounded values of b_t.

Sources: United States Department of Commerce Bureau of the Census, *Historical Statistics of the United States Colonial Times to 1970*, Bicentennial Edition, Part 1, Series E 1 (GNP price deflator) and D 817 (railroad hourly earnings), pp. 197 and 170 respectively; values of $(K/L)_t$ from Table 4.2.

With these estimates of the values of p_L/p_K and of $(K/L)^*$ for each year, we can estimate a value of the production function parameter b for each year. These estimates are given in the second-last column of Table 4.3.

We need now to put a time subscript on the parameter b because its value varies from year to year. We can see that while the value of b was very nearly constant between 1948 and 1952, with $b_{1948} = 0.184814$ and $b_{1952} = 0.183969$, the value of b increased steadily after 1952 to $b_{1957} = 0.197401$, $b_{1962} = 0.226309$, and $b_{1967} = 0.238714$. Over the period as a whole the partial elasticity of output with respect to capital input increased by 29 per cent, so the increase is quite considerable. (Because the parameter b varies according to the year, we subscript b according to the relevant year; when we use b as an exponent, printing and reading a subscript to a superscript is difficult and so we shall write the superscript as $b\{t\}$.)

This increase in b is evidence of the capital-using/labour-saving non-neutral nature of technological change in the railroad industry.

We may estimate a value of Z'_t from the labour productivity form of the production function. Value added at constant 1958 prices, V, is used as a measure of output. This is because the railroad industry produces two main types of quantity of output: passenger-miles (for passengers) and ton-miles (for freight). One way of measuring the 'volume' of output in such a case is to use value added at constant prices. Using V, we may derive a formula for Z'_t from the labour productivity form of the production function as:

$$Z'_t \quad = \quad (V/L)_t/(K/L)_t^{\,b\{t\}}$$

Values of (V/L) and (K/L) for each year are given in Table 4.2, second-last column and last column respectively. The estimated values of Z' for each year are given in the last column of Table 4.3.

The values of Z'_t increase steadily over the years, from 1.820763 in 1948 to 3.188893 in 1967, an increase over the period as a whole of three-quarters, or by 3 per cent per annum compound.

The substantial increases in both b and Z' demonstrate the considerable impact of the technological change which occurred in the railroad industry during these years.

4.4 Unit isoquant analysis of dieselization

Technological change economizes on both factor inputs (at the same scale of annual output). However, in the case of dieselization, we must also take into account the capital-using/labour-saving nature of the technological change, because this means that capital–labour substitution took place as an integral part of the technological change. This would have affected the behaviour of the unit capital requirement in order to achieve these non-neutral reductions in unit labour requirement.

This may be shown by presenting the unit isoquant form of the production function. In these unit isoquants, unit capital requirement and unit labour requirement are measured with respect to annual value added at constant prices, V, in the place of the annual (physical) quantity of output, Q.

Using subscripted values of the parameters (because the unit isoquant shifts according to the (changing) values of these parameters), the formula for the unit isoquant is (Chapter 2 and still keeping to the restriction $a + b = 1$):

$$(K/V) \quad = \quad Z'_{t}{}^{-1/b\{t\}} \, (L/V)^{((b-1)/b)\{t\}}$$

Using the set of unit labour requirements given in the first column of Table 4.2, we may calculate the corresponding values of each unit capital requirement in the unit isoquant for each year.

The results are given in Table 4.4 and the unit isoquants are graphed (in part) in Figure 4.1. The actual position of the railroad industry in each year is given in bold in Table 4.4 and is correspondingly indicated in Figure 4.1.

The unit isoquant shifts steadily downwards and to the left over the years, thus considerably reducing unit labour requirement. This demonstrates the impact of technological change.

In Figure 4.1, the capital–labour ratio may be represented by the slope of a ray from the origin because the value of any such slope is $(K/V)/(L/V) = (K/L)$. The capital–labour ratio actually obtaining in each year is drawn in Figure 4.1.

We can see how the capital–labour ratio consistently increases, so affecting the behaviour of unit capital requirement which increases slightly then decreases. This increase in the capital-intensity of production is due partly to the capital-using/labour-saving non-neutral nature of technological change in the railroad industry, and partly to mechanization in response to the increase in the factor price ratio.

Table 4.4
Unit isoquant form of production function for the United States railroad industry

Unit labour requirement, labour-hours per $ of 1958 price value added, L/V	Unit capital requirement,[a] 1958 $ of capital installed per $ of 1958 price value added per annum, K/V, when:				
	(1948) $Z' = 1.820763$; $b = 0.184814$	(1952) $Z' = 2.070320$; $b = 0.183969$	(1957) $Z' = 2.192491$; $b = 0.197401$	(1962) $Z' = 2.511584$; $b = 0.226309$	(1967) $Z' = 3.188893$; $b = 0.238714$
0.333740	**4.9429**	2.4896	1.6241	0.7279	0.2571
0.281386	10.4918	**5.3069**	3.2502	1.3044	0.4430
0.246030	18.9700	9.6274	**5.6105**	2.0644	0.6798
0.184291	67.8511	34.6831	18.1636	**5.5435**	1.7084
0.134614	271.1798	139.7039	65.1386	16.2236	**4.6518**

(*a*) Calculated from formula for unit isoquant:

$$(K/V) = Z'^{-1/b} (L/V)^{(b-1)/b}$$

(see Chapter 2). The actual unit capital requirement in each unit isoquant is shown in bold.

Source: Tables 4.2 and 4.3.

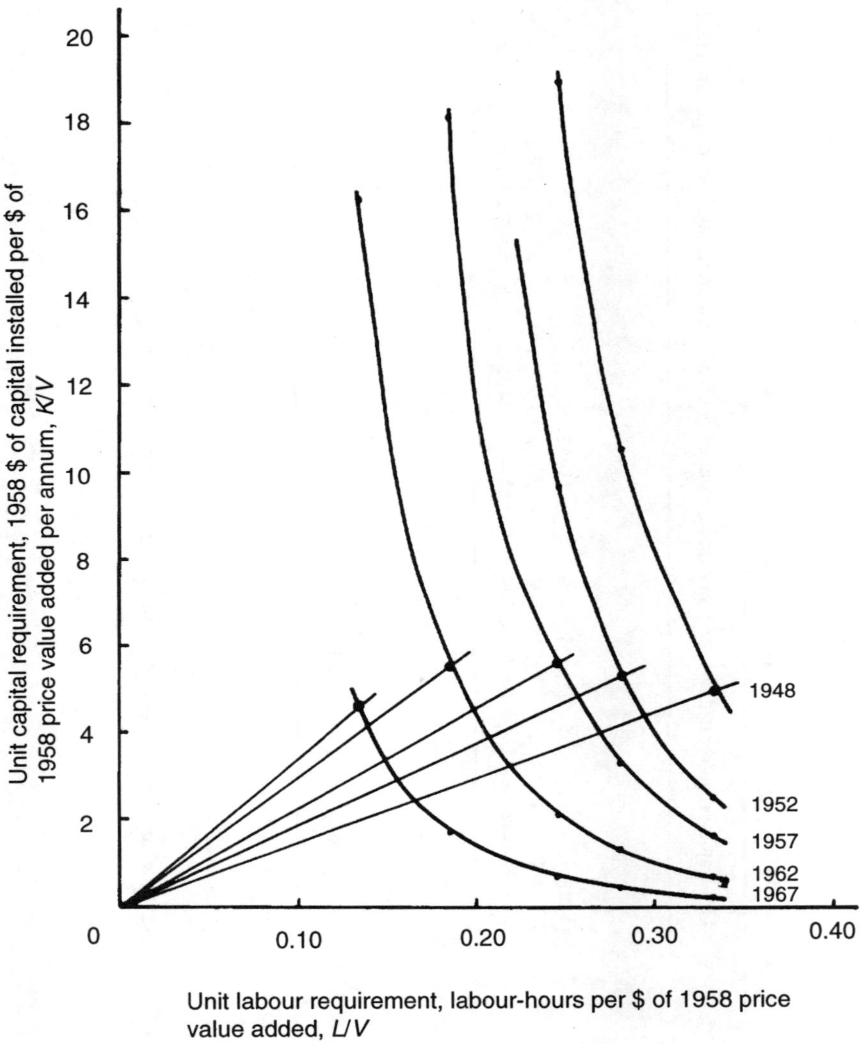

Figure 4.1 Unit isoquants in the United States railroad industry

4.5 The labour productivity production function analysis of dieselization

We may analyse this process of capital–labour substitution by using the labour productivity form of the production function. With value added at constant prices, *V*, as a measure of the quantity of output, the labour productivity form of the production function is:

$$(V/L) \quad = \quad Z'_t\,(K/L)^{b\{t\}}$$

Using the set of values of (*K*/*L*) from Table 4.2, we may calculate the corresponding values of labour productivity, *V*/*L*, according to the labour productivity form of the production function.

The results are given in Table 4.5 and the labour productivity production functions are graphed in Figure 4.2. The actual position of the railroad industry in each year is given in bold in Table 4.5 and is correspondingly indicated in Figure 4.2.

The labour productivity form of the production function shifts steadily upwards over the years, thus considerably raising labour productivity at any given capital–labour ratio. This demonstrates the impact of technological change.

However, there is also movement along the production function due to capital–labour substitution, so that the railroad industry has a considerably greater capital–labour ratio in 1967 than in 1948.

Because of the increase in real average hourly earnings and thence in the factor price ratio (Table 4.3), we know that some of this increase in the capital–labour ratio must have been due to additional mechanization caused by the increase in the factor price ratio.

To what extent was the increase in productivity due to non-neutral technological change and to what extent was the increase in productivity due to the additional mechanization occurring in response to the change in the factor price ratio?

To answer this question we shall take the entire period 1948 to 1967. (The following analysis could be repeated for each of the sub-periods.)

Over the entire period 1948 to 1967, labour productivity increased by a ratio of 2.47923 (= \$7.428632 per labour-hour/\$2.996344 per labour-hour, all at constant 1958 prices – see Table 4.2). This increase is represented in Figure 4.2 by the shift from A to F.

But we can see in Figure 4.2 that, in the first place, this increase involved

Table 4.5
Labour productivity form of the production function for the United States railroad industry

Capital–labour ratio, 1958 $ of capital installed per labour-hour used per annum, K/L	Labour productivity,[a] $ of 1958 price value added per labour-hour, V/L, when:				
	(1948) $Z' = 1.820763$; $b = 0.184814$	(1952) $Z' = 2.070320$; $b = 0.183969$	(1957) $Z' = 2.192491$; $b = 0.197401$	(1962) $Z' = 2.511584$; $b = 0.226309$	(1967) $Z' = 3.188893$; $b = 0.238714$
14.810503	**2.996**	3.399	3.733	4.622	6.068
18.859958	3.133	**3.554**	3.915	4.882	6.429
22.804237	3.245	3.680	**4.065**	5.097	6.727
30.079977	3.416	3.873	4.293	**5.426**	7.187
34.556405	3.504	3.973	4.412	5.599	**7.429**

(*a*) Calculated from formula for labour productivity form of production function (with the restriction $a + b = 1$):

$$(V/L) \quad = \quad Z' \, (K/L)^b$$

The actual labour productivity in each production function is shown in bold.

Source: Tables 4.2 and 4.3.

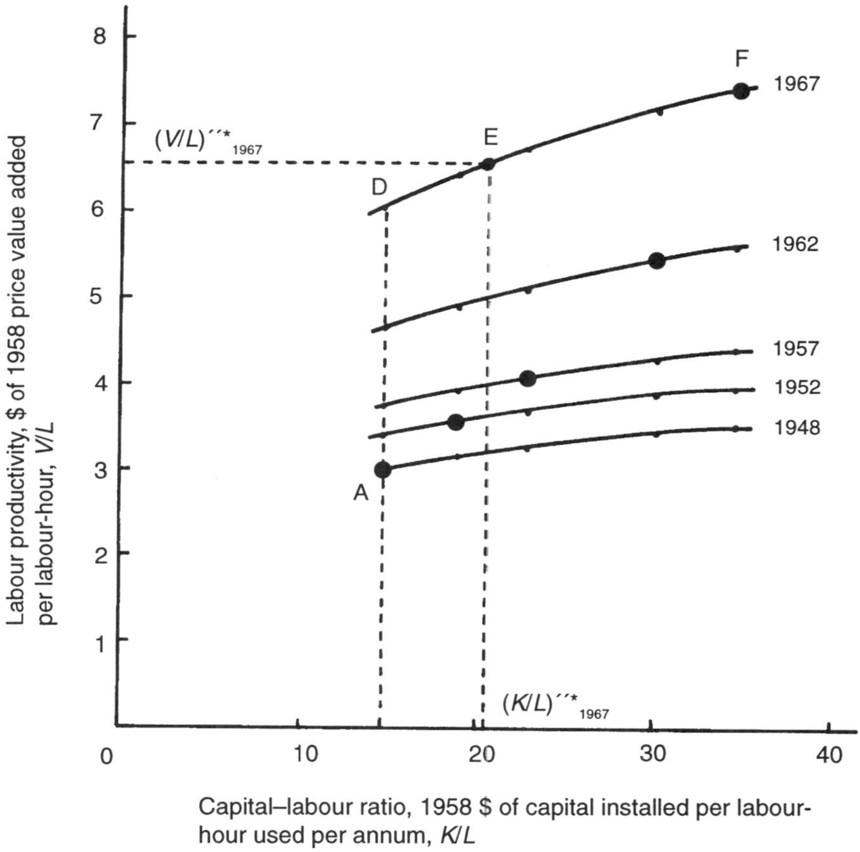

Figure 4.2 Labour productivity production functions in the United States railroad industry

a shift in labour productivity at the same (1948) capital–labour ratio from A to D. This is the impact of the non-neutral shift in the production function (Chapter 3).

In the second place, there was then a movement along the production function from D to F. Some part of this movement will have been due to capital–labour substitution arising because of the increase in the partial elasticity of output with respect to capital input from 0.184814 in 1948 to

0.238714 in 1967 (Table 4.3). In other words, this part of the capital–labour substitution is due solely to the non-neutral nature of technological change in the railroad industry. This part is the non-neutral movement along the (new) production function.

But another part of this movement will have been due to capital–labour substitution or mechanization arising because of the increase in the factor price ratio from 65.326633 in 1948 to 110.204082 in 1967 (Table 4.3). In other words, this part of the capital–labour substitution is due solely to the attempt to reach minimum cost positions of production as these changed with the factor price ratio.

Using the technique explained in the last section of Chapter 3, we may ascertain the extent of the non-neutral movement along the production function and, separately, the extent of mechanization in response to the increased factor price ratio.

We calculate an hypothetical capital–labour ratio as that ratio would have been, at the minimum cost position, after the non-neutral technological change but without a change in the factor price ratio. Denote this hypothetical capital–labour ratio as $(K/L)''*_{1967}$. Then (with data from Table 4.3):

$$(K/L)''*_{1967} = (p_L/p_K)_{1948} / ((1-b)/b)_{1967}$$

$$= 65.326633 / ((1 - 0.238714)/0.238714)$$

$$= \$20.484262 \text{ per labour-hour per annum (constant 1958 prices)}$$

According to the labour productivity form of the production function, labour productivity for this hypothetical capital–labour ratio would have been:

$$(V/L)''*_{1967} = Z'_{1967}\,(K/L)''*_{1967}^{\;0.238714}$$

$$= 3.188893 \times 20.484262^{0.238714}$$

$$= \$6.556836 \text{ per labour-hour (constant 1958 prices)}$$

In Figure 4.2 the hypothetical capital–labour ratio, $(K/L)''*_{1967}$, may be marked on the horizontal axis and the corresponding labour productivity, $(V/L)''*_{1967}$ may then be marked as point E on the 1967 labour productivity form of the production function.

The hypothetical labour productivity represents an increase by a factor of 2.18828 (= \$6.556836 per labour-hour/\$2.996344 per labour-hour), or an increase of 118.8 per cent over nineteen years. This is the measure of the total impact of non-neutral technological change.

We can analyse the components of this technological change using the previous chapter's analytic formula for labour productivity difference:

$$\frac{Z_{1967}}{Z_{1948}} \times (K/L)^{*}_{1948}{}^{b\{1967\}-b\{1948\}} \times \left(\frac{(K/L)''^{*}_{1967}}{(K/L)^{*}_{1948}}\right)^{b\{1967\}}$$

$$= \frac{3.188893}{1.820763} \times 14.810503^{0.238714-0.184814} \times \left(\frac{20.484262}{14.810503}\right)^{0.238714}$$

$$= 1.751405 \times 1.156362 \times 1.080496$$

(1) Proportionate impact of neutral shift in production function (not shown in Figure 4.2)	(2) Proportionate impact of exponent change on shift in production function (not shown in Figure 4.2)	(3) Proportionate impact of non-neutral movement along new production function; from D to E in Figure 4.2

(4) Proportionate impact of non-neutral shift in production function; from A to D in Figure 4.2 [combining (1) and (2)]

$$= 2.025257 \times 1.080496$$

(5) Proportionate impact of capital-using/labour-saving non-neutral technological change; from A to E in Figure 4.2 [combining (4) and (3)]

$$= 2.18828$$

Thus the increase in labour productivity between A and E comprises a 102.5257 per cent non-neutral shift in the production function and an 8.0496 per cent non-neutral movement along the new production function in response to the increased partial elasticity of output with respect to the capital input.

We may estimate the impact on labour productivity of the additional mechanization induced by the increase in the factor price ratio as:

$$\left(\frac{(K/L)^*_{1967}}{(K/L)''^*_{1967}}\right)^{b\{1967\}} = \left(\frac{34.556405}{20.484262}\right)^{0.238714}$$

$$= 1.132958$$

This 13.2958 per cent increase in productivity is illustrated by the movement from E to F in Figure 4.2.

Consequently, the overall increase in labour productivity in the United States railroad industry between 1948 and 1967 was 2.18828 × 1.132958 = 2.47923.

Using these ratio increases, Figure 4.3 depicts, in the form of a cumulative bar chart, the structure of the proportionate increase in labour productivity

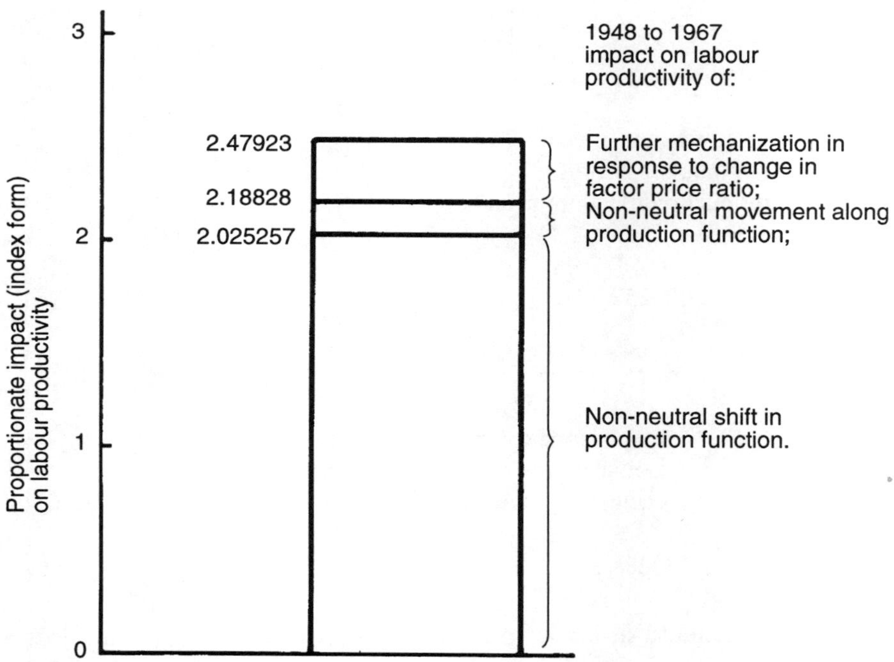

Figure 4.3 Analysis of labour productivity change in the United States railroad industry, 1948 to 1967

due to non-neutral technological change and mechanization in response to the rise in the factor price ratio.

We can immediately see that the major part of the productivity change was due to the non-neutral shift in the production function, and that the additional mechanization induced by the change in the factor price ratio was quantitatively more significant than the mechanization in response to the change in the exponent b.

However, what matters for the whole of such a change in labour productivity is the change in unit labour and capital cost combined. What happened to unit labour and capital cost combined?

4.6 The change in unit labour and capital cost combined

To what extent did all these changes – the non-neutral technological change and the mechanization – reduce or contain the unit labour and capital cost combined?

A formula was developed in the preceding chapter (p. 78) to answer this question in terms of the production function parameters, and for the case where there was a change in the factor price ratio. Where v^*_{1948} denotes unit labour and capital cost combined in 1948, but measured at 1958 prices, and v'^*_{1967} denotes unit labour and capital cost combined in 1967, also at 1958 prices, then, suitably amending the notation and inserting the requisite values from Table 4.3 together with the value $p_K = 0.025$ per annum:

$$\frac{v'^*_{1967}}{v^*_{1948}} = \frac{Z'_{1948}}{Z'_{1967}} \times \frac{(p_L/p_K)_{1967}^{1-b\{1967\}}}{(p_L/p_K)_{1948}^{1-b\{1948\}}}$$

$$\times \frac{\left(\dfrac{1-b_{1967}}{b_{1967}}\right)^{b\{1967\}} + \left(\dfrac{1-b_{1967}}{b_{1967}}\right)^{b\{1967\}-1}}{\left(\dfrac{1-b_{1948}}{b_{1948}}\right)^{b\{1948\}} + \left(\dfrac{1-b_{1948}}{b_{1948}}\right)^{b\{1948\}-1}}$$

$$= \frac{1.820763}{3.188893} \times \frac{(2.755102/0.025)^{1-0.238714}}{(1.633166/0.025)^{1-0.184814}}$$

$$\times \frac{\left(\frac{1-0.238714}{0.238714}\right)^{0.238714} + \left(\frac{1-0.238714}{0.238714}\right)^{0.238714-1}}{\left(\frac{1-0.184814}{0.184814}\right)^{0.184814} + \left(\frac{1-0.184814}{0.184814}\right)^{0.184814-1}}$$

$$= 0.570970 \times 1.188667 \times 1.073557$$

$$= 0.72862$$

Thus, working in terms of constant 1958 prices but allowing for the increase in real average hourly earnings which occurred, the United States railroad industry had in 1967 a (real) unit labour and capital cost combined which was about 27 per cent lower than that in 1948 because of the impact of both capital-using/labour-saving non-neutral technological change and also the additional capital–labour substitution in response to the increased factor price ratio.

The production function approach thus allows us to summarize, and thereby to comprehend, the impact of technological change and of additional mechanization both on labour productivity and on unit labour and capital cost combined.

5 Unit Total Cost and Quality under Technological Change

5.1 Unit total cost

We have seen how technological change increases labour productivity and so reduces unit labour and capital cost combined. But this is not all that technological change is capable of doing.

Technological change is capable of reducing unit materials requirement and unit fuel requirement. Because the cost of materials and fuel tend to be the greater proportion of the total resource costs of production, technological change is thus capable of having a very great impact on both unit total cost and product price. This makes technological change a much more 'powerful' force than any other. But we must have a proper understanding of this saving in materials and fuel.

The laws of physics tell us that we cannot get something for nothing. So what generally happens when either unit materials requirement or unit fuel requirement is reduced is that such a reduction must, according to the laws of physics, come about because of a reduction in 'waste'. That is, some of the materials or fuel input which was being fed into the process was not being effectively used or converted into whatever was wanted but remained in or as unserviceable 'waste' or a low-value by-product.

Technological change often results in, and research and development may have been specifically directed towards, a reduction in waste and hence an increase in output per unit of materials input and/or per unit of fuel input. This has the surface appearance of being 'something for nothing' but is rather (and must be) a reduction in waste.

For example, thermal efficiency is measured as the fraction of heat, used as an 'input', that is converted to effective 'output', seen in mechanical terms such as pulling power or rotatory motion. The external combustion steam engine has a very low thermal efficiency of about 5 per cent, meaning, conversely, that a great deal of the heat 'input' was simply wasted or not used effectively. By contrast, the internal combustion engine has a much greater thermal efficiency. The spark-ignition internal combustion engine has a thermal efficiency of about 20 per cent, meaning that it converts about 20 per cent of the energy in the gasoline into mechanical work; the compression-ignition (diesel) engine has a thermal efficiency of about 30 per cent.

The thermal efficiency fraction is an energy transfer ratio between 'work in' (the denominator) and 'work done' (the numerator). Expressed in this way, thermal efficiency is always well below 1 and thermal efficiency can never equal 1. This is by reason of the Second Law of Thermodynamics, which effectively denies the possibility of a perpetual motion machine (where 'work done' would equal or exceed 'work in').

For an internal combustion piston engine, thermal efficiency may be determined according to the empirical formula: $[1 - 1/(\text{Compression ratio})^{0.4}]$. The compression ratio is the ratio of the maximum to the minimum volume above the piston crown in the cylinder of an internal combustion engine.

Hence the greater the compression ratio, the greater is thermal efficiency. If the compression ratio can be increased, more 'work' or 'energy' can be obtained from the fuel. Within the limits dictated by the need for safety, a high compression ratio is required for thermal efficiency in the use of fuel in an internal combustion piston engine.

Accordingly, the diesel engine, with its higher compression ratio, has a greater thermal efficiency than the spark-ignition engine with its lower compression ratio. The converse of thermal efficiency is what might be called thermal waste, and greater thermal efficiency simply means less thermal waste.

Thermal efficiency is a concept related to, but distinct from, fuel productivity. Fuel productivity is a concept which should be understood as a ratio between a physical quantity of a flow of commercially saleable output per period (say, ton-miles per annum), in the numerator, and a related physical quantity of fuel used during the same period, in the denominator.

To illustrate this, consider the increase in fuel productivity, and the converse reduction in unit fuel requirement, resulting from dieselization in the inter-city railroad industry in the United States. The total quantity of fuel input during a period such as a year can be measured in terms of British Thermal Units, or BTUs. (A BTU is the amount of heat required to raise the temperature of 1 pound of water through one degree Fahrenheit when the water is at or near its point of maximum density at 39.1°F. Each energy source, measured by weight or volume, can be converted to a BTU equivalent. For example, one ton of anthracite coal converts to 25,400 BTUs and a barrel of diesel distillate converts to 5,825 BTUs (United States Department of Labor Bureau of Labor Statistics, *Railroad Technology and Manpower in the 1970's*, Bulletin

1717, p. 81). Nowadays energy is measured in joules and 1 BTU = 1055 joules.) Using BTUs enables us to aggregate together the fuel use of anthracite coal and the fuel use of diesel distillate.

In this illustration, the relevant measure of output is the total gross ton-miles pulled per annum, or 'work done' (these 'gross ton-miles' relate to freight and passengers plus the weight of the railway carriages and freight cars).

The following data from Bulletin 1717 (p. 82) show the remarkable increase in fuel productivity and the converse reduction in unit fuel requirement resulting from dieselization:

United States railroad industry, index, 1948 = 100

	Total fuel use, BTUs per annum	*Gross ton-miles per annum*	*Gross ton-miles per BTU, fuel productivity*	*Fuel use per gross ton-mile, unit fuel requirement*
1948	100	100	100	100
1957	21	91	433	24
1962	17	84	494	20
1966	19	96	505	20

The change-over from steam locomotives to diesel locomotives meant that almost the same annual quantity of work could be done with about 80 per cent less fuel (heat) input. This remarkable change was basically due to the reduction in waste in the use of fuel.

It is this view of 'reduction in waste' which we should generally have in regard to reduced unit materials requirement or reduced unit fuel requirement resulting from process innovation. Obviously, such a reduction in waste will reduce unit total cost.

Another important point may be made here in regard to process innovation. I call it the principle of the importance of 'a small decrease in a big cost' as against 'a big decrease in a small cost'.

Generally speaking, materials cost will have the largest share in total cost. Fuel cost may also have a large share in total cost. Accordingly, a 'small' proportionate decrease in such costs (for any given quantity of output per period) is capable of having a greater impact on unit total cost than a 'big' proportionate decrease in other costs which have a small share in total cost.

To illustrate arithmetically, if materials (or fuel) make up 70 per cent of total cost, then a 'small' 4 per cent decrease in materials (or fuel) input (for the same quantity of output) reduces unit total cost by 2.8 per cent. If labour cost makes up 30 per cent of total cost, then a 'big' 9 per cent decrease in labour cost (for the same quantity of output) reduces unit total cost by 2.7 per cent. The 'small' decrease in the 'big' cost has the same effect as the 'big' decrease in the 'small' cost. If a 'small' proportionate decrease is more readily obtained than a 'big' proportionate decrease, this means that the 'small' decrease in the 'big' cost is more important than the 'big' decrease in the 'small' cost.

Because unit total cost is what ultimately matters, research and development directed towards economizing on the inputs which make up a large share of total cost may be more rewarding than concentrating the R&D effort on inputs which have only a small share in total cost.

Putting on one side the (considerable) costs of research and development, materials technological change may more nearly provide 'something for nothing', especially when new materials enable the same things to be done which formerly required greater quantities of different or more expensive materials. Sometimes technological change will permit the substitution of a less expensive material (or a more readily worked material) for a more expensive material to accomplish the same task, perhaps at a greatly increased capacity and/or in a greatly improved manner. A good example of this would be the substitution of optical fibre cables for copper wire cables in the carrying of signals or the substitution of the transistor for the valve. Such 'new materials' technological change is capable of effecting very considerable reductions in unit materials cost.

Reduction in waste may come about because of a product innovation. For example, the substitution of a micro-wave oven for a conventional oven or hot-plate, and the consequent reduction in household fuel bills for cooking, should basically be seen in terms of greater thermal efficiency or reduction in wasted heat.

Another important point may suitably be made here in regard to product innovation and process innovation. It is the principle of what I call 'interdependence and complementarity'.

By 'interdependence' I mean the mutual, two-way, technical dependence of one product or service upon another product or service, in that each is technically 'no use' without the other. By 'complementarity' I mean that an increase in the demand for one product is associated with, and promotes the demand for, the other product.

Interdependence and complementarity apply both to the initial product innovations and also to subsequent quality improvements in the product. New-plant and in-plant technological change can also be subject to interdependence and complementarity.

In product innovation, it is often the case that development of, or improvement in, product or service A is a prerequisite for development of, or improvement in, product or service B (interdependence), which in turn leads both to increased demand for product or service A (complementarity) and also to further improvement in product or service A (interdependence).

Subsequently in this chapter, we shall see an example of interdependence and complementarity between improved quality gasoline and improved automobile engines. During the 1920s, gasoline was improved in quality and the compression ratio of automobile engines was then increased to take advantage of the higher-quality gasoline. A higher compression ratio improved the thermal efficiency of the spark-ignition internal combustion engine. As the compression ratio of automobile engines was further increased, the petroleum-refining industry attempted (successfully) to produce better quality gasoline to suit. The improvements in the two products (gasoline and automobile engines) were interdependent and complementary.

This brings us to the subject of quality improvement.

5.2 Quality

Surveys of research and development and/or of innovation usually find that 'cost saving' is the most frequently mentioned objective of R&D and innovation. However, in such surveys, 'improving quality' usually ranks almost equally with 'cost saving' as an objective, and may even exceed it. And, indeed, quality improvement is a very important objective of technological change.

Process innovation in the form of improved quality control may reduce waste by reducing the proportion of finished products which cannot be sold due to defects of manufacture. This is capable of having a significant impact on unit total cost.

If we define, or redefine, Q to be the saleable output (free of defects) then simply by reducing the proportion of rejected items of output, technological change increases Q for the same quantity of all inputs, thereby reducing all unit input requirements and hence unit total cost (that is, total cost per unit of saleable product).

This is another aspect of reduction in waste which we must keep in mind when considering technological change.

But there is more to quality than simply a reduction in waste. A common feature of quality improvement is what I call the principle of 'increased service'.

In general, products are not desired as objects in themselves but are wanted for the service(s) which they provide. Very broadly, such services may be called 'features' of the product and improved features are a common result of technological change.

Consider the case of the tyre industry in the United States which experienced considerable technological change in the 1920s. Process innovations in the manufacture of tyres led to increased labour productivity. But this was far from being the most important part of the change, because the average 'life' of a tyre more or less doubled from 8,000 miles (about 13,000 kilometres) per tyre to 15,000 miles (about 24,000 kilometres) per tyre. Although better roads contributed to this, the tyres had also to cope with heavier and faster automobiles. The main reasons for the increased mileage-life of a tyre were improvements in the quality of the materials and in the design of the tyre.

We have the following data for the United States tyre industry (United States Department of Labor Bureau of Labor Statistics, *Labor Productivity in the Automobile Tire Industry*, Bulletin No. 585, July 1933, table 2, p. 3, and Work Projects Administration National Research Project, *Industrial Research and Changing Technology* Report No. M-4, January 1940, table 7, p. 34):

	1923	*1929*
Number of employees in tyre industry	73,963	83,263
Number of tyres produced per annum, thousands	46,369	70,189
Tyres produced per employee per annum	626.9	843.0
Average mileage-life of a tyre, miles	8,000	15,000
Tyre-miles of service produced per employee per annum	5,015,200	12,645,000

If labour productivity is conventionally measured as quantity of the product per unit of labour input, then labour productivity increased by 34.5 per cent between 1923 and 1929. But is this the appropriate view to take?

We see that the average life of a tyre more or less doubled over this period, and if we take a broader view of the services which purchasers of tyres want, then labour productivity, measured by the total tyre-miles of service produced per employee per annum, increased by a factor of 2.52, or by 152 per cent!

It is this sort of change, not simply number of objects (tyres) produced per employee per annum, which is important in determining changes in standards of living.

5.3 Technological change and reduced unit cost in pulp making

We may illustrate the impact of technological change on unit total cost by considering capital-using/labour-saving non-neutral technological change in a plant making pulp as part of the process of making paper. The technological change involved the change from a batch process to a continuous process for pulp making. Changing from batch production to continuous production is a common sort of technological change, and so it is useful to study an example of this. The information is taken from a valuable report by the United States Department of Labor Bureau of Labor Statistics, *Impact of Technological Change and Automation in the Pulp and Paper Industry*, Bulletin No. 1347, October 1962, hereafter cited as Bulletin 1347.

Paper is usually made by converting wood into pulp and processing the pulp into paper. In order to make the pulp, logs must be chipped and the woodchips are then 'cooked' under pressure with chemicals in 'cooking liquor' to produce a pulp of cellulose. This cellulose pulp is the material from which paper is made.

In Plant B, the subject of one of the studies in Bulletin 1347, pulping was done in batches in large steel tanks called 'digesters' and was then washed in 'diffuser tanks' and refined before being sent to the paper mill.

The batch method required employees manually to fill the digesters with woodchips and cooking liquor, to discharge the digesters into diffuser tanks for washing, and then to discharge the diffuser tanks. Under the batch process there was considerable manual and repetitive 'stop–start' manipulation of machinery (Bulletin 1347, pp. 51–3 and 57–61). The batch method necessarily wasted heat in repeated cooling and heating of digesters between batches.

Furthermore, the batch process did not extract the maximum possible quantity of cellulose from any given quantity of woodchips; in other words,

Table 5.1
Impact on unit cost of technological change in pulp plant of Plant B, United States, ca. 1955

		(1) Former batch production technology	(2) New continuous process technology	(3) = (2) – (1) Absolute change	(4) = ((3)/(1)) 100 Percentage change
Units of measurement as given (except for percentage change)					
(1)	*Value of plant and equipment installed (at date of purchase), $, *K*	1,500,000	1,085,000	–415,000	–28
(2)	*Number of workers	12	8	–4	–33
(3)	Total annual input of labour-hours,[a] *L*	26,580	17,720	–8,860	–33
(4)	Capital–labour ratio, $ of capital installed per each labour-hour used per annum, *K*/*L*	56.43	61.23	4.80	9
(5)	*Tons of pulp per day produced by plant	119	126[b]	7	6
(6)	Tons of pulp produced per annum,[c] *Q*	29,750	31,500	1,750	6
(7)	Unit labour requirement, labour-hours per ton, *L*/*Q*	0.8934	0.5625	–0.3309	–37
(8)	Unit capital requirement, $ of capital installed per ton per annum, *K*/*Q*	50.4202	34.4444	–15.9758	–32
$ per ton (except for percentage change)					
(9)	Unit labour cost,[d] p_L (L/Q)	1.8538	1.1694	–0.6844	–37
(10)	Unit capital cost,[e] p_K (K/Q)	5.0420	3.4444	–1.5976	–32
(11)	Unit labour and capital cost combined, *v*	6.8958	4.6138	–2.2820	–33
(12)	Unit woodchip (materials) cost[f]	12.1669	11.0608	–1.1061	–9
(13)	Unit chemicals (materials) cost[g]	2.2800	2.2800	0	0
(14)	Unit other (fuel) cost[h]	5.0074	3.3383	–1.6691	–33

cont.

Table 5.1
Impact on unit cost of technological change in pulp plant of Plant B, United States, ca. 1955 (continued)

		(1) Former batch production technology	(2) New continuous process technology	(3) = (2) – (1) Absolute change	(4) = ((3)/(1)) 100 Percentage change
(15)	Unit materials cost, (12) + (13) + (14), *UMC*	19.4543	16.6791	–2.7752	–14
(16)	Unit total cost, (11) + (15), $v + UMC = u$	26.3501	21.2929	–5.0572	–19

* Denotes data given in report; all other figures are estimates, mostly based on information in Bulletin 1347.

(*a*) Estimates based on average weekly hours for production workers in the pulp, paper and board industry in 1955 of 44.3 (Bulletin 1347, p. 84) and an assumed 50-week working year.

(*b*) Anticipated daily output; actual daily output in the first year after the change was substantially lower at 99 tons, because 'equipment downtime for maintenance was much greater than expected ... Foreign objects in the woodchips, and malfunctions in moving parts caused some costly shutdowns during the first few months of operation' (Bulletin 1347, p. 53).

(*c*) Estimates based on an assumed 250-day working year for the plant, allowing for normal downtime.

(*d*) Estimates based on an average hourly wage rate of $2.075 before the technological change and $2.079 after the technological change (Bulletin 1347, p. 62).

(*e*) Estimate based on an assumed rate of depreciation of 0.10 per annum.

(*f*) Value for former system estimated according to proportionate cost structure given in Bulletin 1347, table 17, p. 55, and ratio to unit labour cost (operating and maintenance); value for new system estimated on basis of text statement that the 'new equipment yields 10 percent more pulp from the same quantity of wood' (Bulletin 1347, p. 53).

(*g*) Value for former system estimated according to proportionate cost structure given in Bulletin 1347, table 17, p. 55, and ratio to unit labour cost; value for new system estimated on assumption of no change.

(*h*) Value for former system estimated according to proportionate cost structure given in Bulletin 1347, table 17, p. 55, and ratio to unit labour cost; I have assumed that the cost structure of table 17 excludes unit capital cost (on the basis of the fact that Bulletin 1347 makes no mention of unit capital cost and the presumption that, had unit capital cost been included, Bulletin 1347 would have contained some statement about the method of its calculation); value for new system estimated on basis of text statement that in the new system 'about a third less steam per ton of pulp produced is used' (Bulletin 1347, p. 53); this also assumes that 'other cost' is entirely (or at least very largely) fuel and power cost.

Source: United States Department of Labor Bureau of Labor Statistics, *Impact of Technological Change and Automation in the Pulp and Paper Industry*, Bulletin No. 1347, October 1962, tables 15 and 16, pp. 53 and 54.

there was some non-extraction 'wastage' of the cellulose present in the woodchips.

When the time came to replace the digesters and diffuser tanks, a decision was taken to switch to a new technology involving continuous production.

In the system of continuous production, woodchips, together with the required 'cooking liquor', were fed continuously into a 'continuous digester' and then continuously and automatically discharged to a continuous and automatic washing cycle.

The continuous nature of the operation saved considerably on heat input, and about one-third less steam per ton of pulp was used in the cooking process, thus reducing unit fuel cost.

In the continuous washing cycle, the pulp was 'defibred', a process not done under the batch system. This increased by 10 per cent the quantity of cellulose pulp obtained from any given quantity of woodchips. In other words, less cellulose was wasted, due to non-extraction, under the continuous process. Unit materials requirement was thereby reduced, leading to a reduction in unit materials cost.

Throughout the process, labour-time was saved. Continuous conveying of the materials eliminated the labour required for handling batches of materials. Effectively, unskilled workers doing manual labour were replaced by semi-skilled workers monitoring a mechanized process (Bulletin 1347, pp. 25–7 and 51–3).

Table 5.1 shows the results of this technological change.

First, the new system cost about 28 per cent less than did the former system (at date of purchase); that is, \$1.085 million for the continuous process equipment as opposed to \$1.5 million for the batch production equipment.

Second, the number of workers in the pulp plant was reduced by one-third from 12 to 8. The estimated quantity of labour-hours used per annum was likewise reduced by 33 per cent.

Together this meant an increase of 9 per cent in the capital–labour ratio, so this is prima facie evidence of capital-using/labour-saving non-neutral technological change.

Third, the quantity of annual output produced was not much increased so this indicates that economies of scale was not a feature of this particular change. Only one part of an integrated plant was being changed, so, unless the scale of the entire plant were being increased, there was no need to increase the scale of output from the pulp plant.

Fourth, the result of all this was that unit labour requirement was decreased by 37 per cent and unit capital requirement was decreased by 32 per cent, with corresponding decreases in unit labour cost and unit capital cost, leading to a 33 per cent decrease in unit labour and capital cost combined.

So far we have been on familiar ground: technological change leading to a decrease in unit factor input requirements and in unit factor cost.

The new and important feature of Table 5.1 is its illustration of the impact of technological change on unit materials cost and on unit fuel cost.

Fifth, as a consequence of more complete extraction of cellulose from woodchips, unit woodchip (materials) cost decreased by 9 per cent. But because unit woodchip cost was a big proportion, about one-half, of unit total cost, this 'small' reduction in unit woodchip (materials) cost contributed considerably to the reduction in unit total cost. This is an example of the importance of a 'small decrease in a big cost'.

Sixth, as a consequence of eliminating waste of heat, unit fuel cost decreased by 33 per cent.

Finally, as a consequence of all of the foregoing unit total cost decreased by 19 per cent. This demonstrates the power of technological change to reduce unit total cost. Such cost reductions inevitably lead to price reductions and so to increased standards of living in terms of the 'affordability' of products.

5.4 Technological change, improved quality, and product price in petroleum refining

Technological advances in petroleum refining illustrate quality improvements and reduced unit total cost with consequent substantial impact on product price. This case study concerns advances made since 1913 in the refining of crude petroleum into gasoline, middle distillates (kerosene and heating oil), and residual fuel oil. The information is mostly taken from an exceedingly illuminating and valuable study by John Enos, *Petroleum Progress and Profits: A History of Process Innovation* (The MIT Press, 1962), hereafter cited as *Petroleum Progress and Profits*.

The present summary considerably simplifies Enos' detailed account of what is a very complex story by concentrating upon the 'new-plant' technological changes which occurred.

Enos also describes the 'in-plant' technological changes which further improved each new process resulting from new-plant technological change. These in-plant technological changes were quantitatively important: 'Each new ... process, as originally installed, did not represent the perfection of the art. Rather, throughout the economic life of each process, there were a series of [subsequent in-plant] improvements. These ... accumulated improvements tend to contribute *just as much* to technological progress.' 'The evidence from the petroleum refining industry indicates that improving a process [in-plant] contributes *even more* to technological progress than does its initial development.' (First quotation from *Petroleum Progress and Profits*, p. ix; second quotation from J.L. Enos, 'A Measure of the Rate of Technological Progress in the Petroleum Refining Industry', *Journal of Industrial Economics*, June 1958, p. 180; italics added in both quotations.)

Generally, I omit discussion of the important role played by economies of scale. This too is discussed at length by Enos, but detailed consideration of scale economies in petroleum refining would be out of place for present purposes.

The basic data are given in Table 5.2.

Over the period 1913 to 1955, technological progress in petroleum refining has enabled refineries to produce, from a barrel of crude petroleum, a much greater yield of gasoline (or 'petrol'), and also gasoline of better quality.

Quality of gasoline is measured by its octane rating. The rating goes from 0 for a (low-grade) reference fuel called heptane which tends to cause 'knocking' in a standardized four-stroke internal combustion spark-ignition piston engine, to 100 for a (high-grade) reference fuel called trimethylpentane which has very high knock-resistance.

To determine the octane rating of a sample of gasoline, heptane and trimethylpentane are mixed in proportions which produce a blend to match the knock behaviour of the sample. The percentage of trimethylpentane (the high-grade reference fuel) in the blend is then called the octane rating of the sample gasoline.

A higher octane gasoline permits the use of an engine with a higher compression ratio which thus extracts more thermal energy from the fuel, so giving better performance in terms of ton-miles per gallon carried at a standard speed.

The phenomenon known as 'knocking' or 'detonation' in internal combustion spark-ignition piston engines is caused by the self-ignition, or spontaneous combustion, of the last part of the compressed fuel–air mixture in the combustion chamber after the other part of the mixture has been ignited by the spark. The advancing spark-ignited flame front compresses the unignited part of the fuel–air mixture, which then 'detonates' or self-ignites under the compression. This 'detonation' produces a rise in pressure from the two conflicting explosions in the confined space of the combustion chamber. This pressure is heard as a knocking noise from the engine.

Table 5.2
Technological change in the refining of crude petroleum, United States, 1913 to 1955

Process	Date	Octane rating of gasoline	Product yields in gallons per barrel (of 42 US gallons) of crude petroleum used in ('charged to') the process, gallons[a]			Unit petroleum requirement, gallons of crude petroleum required per gallon of gasoline
			Gasoline	Middle distillates (kerosene, heating oil)[b]	Fuel oil	
Crude distillation	1913	n.a.	8.61	6.83	24.63	4.88
Burton batch process[c]	1914	55	10.65	8.49	20.05	3.94
Burton–Clark batch process with bubble tower	1922	72	17.35	4.49	17.05	2.42
Tube and Tank continuous process	1922	72	16.38	15.11	6.30	2.56
Houdry catalytic process, semi-continuous	1939–40	87.7	15.6	25.8	–	2.69
Fluid catalytic process, continuous[d]	1942	95.4	17.6	16.40	0.93	2.39
Fluid catalytic process, continuous, large scale[d]	1955	94.7	24.7	11.3	1.35	1.70

(*a*) 1 US gallon = 3.7854121 litres.

(*b*) During 1913 to 1922, middle distillates were sold as kerosene. From 1936 onwards, they have been sold as heating or furnace oil.

(*c*) Installation with hinged false bottom plates in the still to facilitate cleaning and increase the operating time between cleanings. This was an 'add on', or in-plant technological change, to the original 1913 Burton installation. While improving operating efficiency, the octane rating of the gasoline produced remained the same in plants of 1913 vintage and 1914 vintage.

(*d*) Product yields for the fluid process do not vary with the capacity of the unit (that is, there are no economies of scale to use of *materials* (petroleum) input); there were, however, economies of scale to *factor* inputs.

Source: John Enos, *Petroleum Progress and Profits: A History of Process Innovation* (The MIT Press, 1962), Appendix tables 8a and 8b, p. 299; 'n.a.' means 'not available'.

Knocking is an uncontrolled and violent process (as opposed to the controlled and orderly combustion process which takes place when the fuel–air mixture is properly ignited). Knocking leads to mechanical and heat stresses on the engine, to loss of power, and to reduced efficiency of fuel use. The heat generated in the engine by continued severe knocking can lead to the separate and distinct phenomenon of pre-ignition, which is self-ignition of the whole of the fuel–air mixture before the piston has completed its upward compression stroke and before the spark ignition. Pre-ignition can lead to severe damage to the engine; for instance the seizure and breaking-up of the piston.

Consequently, knocking is highly undesirable and absolutely to be avoided, especially in piston aero-engines! Knocking can readily occur as it is an inherent behavioural tendency of the fuel–air mixture in spark-ignition engines. Any such engine can be made to knock by using the 'wrong' grade of fuel or by altering its compression ratio, and a great deal of research and development has gone into improving both the quality of the fuel and the design and working of engines so as to avoid knocking.

Knocking sets an upper limit to the compression ratio and hence to the performance of an engine. A low-octane gasoline causes knocking at a low compression ratio; therefore a low-octane fuel necessitates a thermally inefficient low compression ratio in the engine to avoid knocking. As the octane rating of the fuel is increased, the compression ratio of the engine can be increased (without causing knocking), and this leads to greater thermal efficiency and possibilities for better mechanical performance.

The first person correctly to explain the phenomenon of knocking (or 'detonation' as he called it), to establish that knocking set an upper limit on the efficiency of the internal combustion piston engine, and to reveal the connection with octane rating was the automotive engineer Sir Harry Ricardo in 1919 (see: H.R. Ricardo, *The High-speed Internal-combustion Engine* (Blackie, Fourth Edition, 1953), Chapter II Detonation and Pre-ignition, Chapter III Efficiency, and Chapter IX Mechanical Efficiency).

Ricardo realized that there were two ways of dealing with the problem: either to use a fuel of higher octane which would not self-ignite; or to design the combustion chamber in such a way as to inhibit knocking. In the years following 1919 both ways were developed to the very great benefit of motorists, to say nothing of increased safety in aircraft. Ricardo thus established the basis, and the direction, for much of the technological change in petroleum refining.

Following Ricardo's explanation of knocking, much research and development in petroleum refining was directed to increasing the octane rating of gasoline because this greatly increased the 'usefulness' of the product. Improved quality gasoline and improved automobile engines were very interdependent and complementary innovations.

Up to about 1913, gasoline was distilled from crude petroleum in a batch process simply by boiling that petroleum at atmospheric pressure and successively recovering the volatile elements by distillation. As shown in Table 5.2, the process of simple distillation in a plant of 1913 vintage produced about 8.61 US gallons of gasoline per barrel (of 42 US gallons) of crude petroleum. (Throughout this section, 'gallons' must be understood as US gallons; 1 US gallon = 3.785 litres; 1 UK (or 'Imperial') gallon = 4.546 litres.)

Since 1861 it had been known that heating crude petroleum under pressure, a process known as 'thermal cracking' or simply 'cracking', produced a greater quantity of gasoline from a given charge of crude petroleum. Furthermore, thermal cracking produced gasoline with a higher octane rating.

In 1913, an employee of the Standard Oil Company (Indiana), Dr William Burton, a university graduate with a PhD in chemistry, patented a batch method of cracking gasoline. As Table 5.2 shows, in a plant of 1914 vintage the Burton process raised the yield of gasoline to 10.65 gallons of gasoline per barrel of crude petroleum. This was an improvement in quantity of about one-quarter on methods previously available.

The improved yield of gasoline from a barrel of crude was accompanied by a reduced yield of fuel oil, but gasoline is a more remunerative product. However, the quality, as measured by the octane rating, of gasoline produced by the Burton process left a lot to be desired.

The original Burton process was itself subsequently improved both by the extension of the tank into a 'bubble tower' where a greater volume of gasoline distillate could be captured at the top of a tower, and by an innovation of the manager of one of the Indiana Standard refineries, Mr Edgar Clark, who cracked the crude petroleum in a set of tubes rather than in a tank. The advantage was that tubes needed to be cleaned only once in 48 hours whereas Burton's tank need cleaning once every 24 hours.

By 1922, the improved Burton–Clark process with bubble towers gave a yield of 17.35 gallons of gasoline per barrel of crude. The octane rating was improved to 72. In less than ten years the thermal cracking process had more than doubled the yield of gasoline and had increased by one-third the quality of the gasoline.

The next innovation was the introduction by the Standard Oil Company (New Jersey) of the Tube and Tank continuous process. In 1918 Jersey Standard induced Clark to come and work for them on the problem of cracking high sulphur Mexican and Californian crude oil (because of their sulphur content and high viscosity, these oils were not suitable for cracking by the Burton process). Clark developed a continuous process thermal cracking system of tubes and tanks operating at higher pressures with steam introduced into the process to reduce the viscosity of the crude petroleum.

As a self-trained practical engineer, rather than a chemist, Clark understood the need for, and the potentialities of, a continuous cracking process, and he deliberately directed his research and development at Jersey Standard to making continuous cracking technically feasible and commercially viable.

By 1922, Clark had obtained from this continuous process about the same yield of gasoline (of the same quality) but a greatly increased quantity of middle distillates. The quantity of residual fuel oil was correspondingly greatly reduced. Clark's Tube and Tank thermal cracking continuous process also permitted operations to be carried out on a larger scale, and this substantially reduced unit capital cost and unit fuel cost.

During this period and subsequently, the manufacturers of motor vehicles improved the performance of vehicles and this required higher octane gasoline. Eugene Houdry was a French engineer with a great interest in automobile racing, and he became greatly concerned with the problem of producing higher octane gasoline. Houdry devoted his research to methods of cracking crude petroleum in the presence of a catalyst which would take up non-volatile carbonaceous material from the crude petroleum as a coating of coke, and at the same time ensure the production of higher octane gasoline. As part of the process the coke deposits can be subsequently burned off to 'regenerate' the catalyst. The whole process is known as 'catalytic cracking'. The catalyst enhances the cracking reactions and promotes the desired chemical reactions, so producing higher octane gasoline. Catalytic cracking (with catalysts other than the one discovered by Houdry) had been known and used prior to Houdry's work but had been very expensive due largely to the high cost of those catalysts. Houdry was effectively the pioneer of commercially viable catalytic cracking.

After a great deal of experimenting with catalysts, Houdry discovered, at three o'clock on an April morning in 1927, a suitable catalyst of oxides of silicon and aluminium. Taking the process to a commercial stage required specialized equipment, and in 1930 Houdry went to the United States where he was able to obtain the funding to continue his research and development.

The Sun Oil Company, which specialized in producing high-quality gasoline, took up Houdry's process and, after solving various difficult production problems, the first Houdry catalytic crackers started production in 1936. With subsequent in-plant improvements, by 1939 this process was producing a reasonable yield of gasoline of high quality, 87.7 octane rating, together with a further improved yield of middle distillates.

The Houdry process had improved the octane rating of gasoline by over one-fifth, and its chief advantage lay in this quality improvement. Subsequent in-plant technological change involved such things as better control of heat transfer and the development of synthetic catalysts in 'pelleted' form which improved the catalytic reaction.

By 1942 there were fourteen Houdry units in operation and during the early years of the Second World War these units produced most of the catalytically cracked aviation gasoline.

However, the Houdry catalytic process was a semi-continuous process: the refining process had to be switched between vessels, with one vessel doing the refining while in the other vessel the catalyst in its 'fixed bed' was being regenerated. Because of having more than one vessel and because of the various switching and control mechanisms, the unit capital requirement was high in the Houdry process. This provided an obvious challenge to petroleum refinery engineers to combine the greater chemical effectiveness of the catalytic cracking process with the lower unit capital requirement of a completely continuous process.

Although each of the Burton process and the Houdry process 'represented significant improvements over the previously existing art, each was in turn superseded within seven years; the main reason for the rapid obsolescence was the lower unit costs achieved in continuous, rather than batch or semicontinuous, operations in any industry processing extremely large volumes of materials' (*Petroleum Progress and Profits*, p. 162).

In 1938 a group of firms led by Standard Oil (New Jersey) organized a co-operative research team of about 1,000 scientists to work on improvements to the catalytic cracking process, stimulated partly by the high royalty payments required on the Houdry process and partly by the intention to develop a completely continuous catalytic cracking process.

The basic idea was to have a moving catalyst which would improve the reaction between catalyst and crude petroleum and reduce the cracking time. A 'fluid bed' of powdered catalyst was found to be technically feasible, and in 1942 the fluid catalytic continuous process began producing gasoline of very high quality (95 octane) with a higher yield of gasoline, 17.6 gallons per barrel of crude.

Subsequent in-plant improvements to the fluid bed process raised the yield of gasoline to 24.7 gallons per barrel. These changes concerned such things as modifications to the basic design, improvements in components, for example reducing corrosion by the use of steel alloys, and better process control.

Importantly, the fluid bed catalytic process was completely continuous, because the powdered catalyst flowed through the crude petroleum in a continuous stream with the oil vapours being produced in a continuous and uninterrupted stream. The powdered catalyst was also regenerated in a continuous process.

Because it needed fewer vessels and less ancillary equipment for carrying out the refining, the fluid catalytic process required less steel for constructing the plant and so had a lower unit capital requirement than the Houdry semi-continuous process.

Furthermore, plants for the fluid catalytic process were built on a larger scale (stimulated by the desperate need, during wartime, not only to increase production but also to economize on scarce steel). It is clear that there were considerable economies of scale in petroleum refining. Enos gives the following figures for the continuous catalytic cracking process: a plant with a crude petroleum charge capacity of 2,000 barrels per day had an acquisition cost of $800,000 (1946 prices), while a plant with a charge capacity of 20,000 barrels per day had an acquisition cost of $4,700,000 (1946 prices – *Petroleum Progress and Profits*, p. 212, note 53). The difference in the acquisition cost of the plant is in the ratio 5.875, while the difference in charge capacity (related to quantity of output) is in the ratio 10. Consequently, the unit capital requirement in the larger plant would be 41.25 per cent less than the unit capital requirement in the smaller plant.

We have traced the story of one of the more remarkable technological developments of this century. This development shows one of the hallmarks of technological change in that unit materials requirement was reduced very substantially: from 4.88 gallons of crude petroleum required per gallon of gasoline in 1913 to 1.70 gallons of crude petroleum required per gallon of gasoline in 1955; a reduction of nearly two-thirds!

We should now look more closely at the economic results of these technological changes.

We concentrate on gasoline but we should also take account of the by-products of the distillation process (middle distillates such as kerosene, and the residual fuel oil) to ascertain the cost of producing gasoline net of the value of by-products. This is because, in accordance with our view of 'waste', increased gasoline production means decreased production of by-products and this must be allowed for in any comprehensive assessment of technological change as it affected the distillation process.

Furthermore, gasoline prices at the refinery will be affected by by-product recovery, and so too will wholesale and retail prices, all of these being subsequently considered because of their importance for standards of living.

For the sake of the illustration, we do not consider in detail the (substantial) contribution of economies of scale which were inextricably mixed up with technological change. It is clear that technological change permitted and encouraged the construction of large-scale plants. However, the reduction in crude petroleum input per unit of gasoline output was clearly due only to technological change, as was the improvement in the quality of gasoline.

Table 5.3 gives the available basic data on unit input requirements under the 1914 thermal cracking Burton batch process and the 1955 fluid catalytic cracking large-scale continuous process, together with a set of 1939 prices, so that we can calculate the reduction in unit cost brought about by technological change. (Note however the very large increase in scale indicated in footnotes (*a*) and (*b*) to the table.)

Table 5.3 also contains data for estimating the impact of by-product recovery on the 'effective' or 'net' unit cost of producing gasoline; that is, the unit cost of gasoline net of the recovery of by-products, because this latter effectively reduces the materials unit cost of gasoline.

In the first place, technological change decreased the unit (raw) material input requirement with respect to the output of gasoline. The unit input requirement for crude petroleum charged to the cracking process decreased from 396 gallons of crude petroleum per 100 gallons of gasoline produced to 170 gallons of crude petroleum per 100 gallons of gasoline produced. This is a very large reduction, and is the sort of thing which can occur only under technological change.

At a 1939 price of $1.20 per barrel of crude, or $0.02857 per gallon of crude, this gave a unit crude petroleum input cost of $11.31 (= $0.02857 × 396 gallons of crude) per 100 gallons of gasoline produced by the 1914 process, and a unit crude petroleum input cost of S4.86 (= $0.02857 × 170 gallons of crude) per 100 gallons of gasoline produced by the 1955 process. This is a very large cost reduction, both in absolute terms and in relative terms.

In the second place, technological change (and economies of scale) decreased the unit process labour requirement from 1.61 process labour-hours per 100 gallons of gasoline produced to 0.02 process labour-hours per 100 gallons of gasoline produced.

At a 1939 price of $1 per labour-hour (for process workers), this gave a unit labour cost of $1.61 per 100 gallons produced by the 1914 process and $0.02 per 100 gallons produced by the 1955 process.

In the third place, technological change decreased the unit energy

Table 5.3
The decrease in unit input requirements, and hence in unit costs (at constant input prices), due to technological change (and economies of scale) in petroleum refining

Input	Input prices, 1939	Unit input requirement per 100 gallons of gasoline: Burton process 1914[(a)]	Unit input requirement per 100 gallons of gasoline: Large-scale fluid process 1955[(b)]	Unit cost per 100 gallons of gasoline, $ at 1939 prices: Burton process 1914	Unit cost per 100 gallons of gasoline, $ at 1939 prices: Large-scale fluid process 1955	Change 1914 to 1955
Crude petroleum	$1.20 per barrel (42 gals) $0.02857 per gallon	396 gals	170 gals	11.31	4.86	–6.45
Process labour	$1 per labour-hour	1.61 labour-hrs	0.02 labour-hrs	1.61	0.02	–1.59
Energy	$0.131 per mn BTUs	8.4 mn BTUs	1.1 mn BTUs	1.1004	0.1441	–0.9563
Capital[(c)]	$0.35 per $1 invested per 100 gals p.a.	$10.26	$1.47	3.6	0.52	–3.08
Gross unit total cost	–	–	–	17.6204	5.5441	–12.0763
Recovery of by-products:	*By-product output price, 1939*	*By-product recovery, in gallons, from input of crude per 100 gallons gasoline*		*Unit recovery per 100 gallons of gasoline, $ at 1939 prices*		
Middle distillate (kerosene)	$0.047 per gallon	80.049[(d)]	45.738[(e)]	3.7623	2.1497	–
Fuel oil	$0.0172 per gallon	189.043[(d)]	5.464[(e)]	3.2515	0.0940	–
Total recovery	–	–	–	7.0138	2.2437	–
Net unit total cost	–	–	–	10.6066	3.3004	–7.3062

(*a*) At 88.5 barrels of crude petroleum throughput per calendar day; 1 barrel = 42 gallons. (*b*) At 36,000 barrels of crude petroleum throughput per calendar day. (*c*) Capital costs include depreciation on construction, maintenance, royalty, and, for the fluid process, the annual cost of the catalyst. Unit capital requirement is per 100 gallons per annum, and is an approximation for illustrative purposes only; the unit capital requirement is considerably affected by increasing returns to scale for 1955 figure. (*d*) Calculated as (396/42) multiplied by product yield from Table 5.2. (*e*) Calculated as (170/42) multiplied by product yield from Table 5.2.

Source: John Enos, *Petroleum Progress and Profits: A History of Process Innovation* (The MIT Press, 1962), table 5, p. 224; Appendix tables 5 and 11; pp. 295 and 303; by-product prices (export price for kerosene and Oklahoma refinery price for fuel oil), Appendix tables 3 and 4, pp. 292–3; unit capital requirement based on author's calculations from *Petroleum Progress and Profits*, table 5, p. 224, and Appendix tables 6, 8a, 8b, 13a, and 13b, pp. 296, 299, 306, and 307.

requirement from 8.4 million BTUs per 100 gallons of gasoline produced to 1.1 million BTUs per 100 gallons of gasoline produced.

At a 1939 price of $0.131 per million BTUs, this gave a unit energy cost of $1.1004 per 100 gallons produced by the 1914 process and $0.1441 per 100 gallons produced by the 1955 process.

In the fourth place, technological change (and significant economies of scale) decreased the unit capital requirement (at constant 1939 prices) from $10.26 of capital installed per 100 gallons of gasoline produced per annum to $1.47 of capital installed per 100 gallons of gasoline produced per annum.

At a constant price, or rate of depreciation, of 0.35 per annum per dollar of capital installed, this gave a unit capital cost (according to Enos' figures) of $3.6 per 100 gallons produced by the 1914 process and $0.52 per 100 gallons produced by the 1955 process.

Pause to note the substantial reductions in (physical) unit input requirements for *all* inputs. This is a distinguishing characteristic of technological change.

Taking the sum of all the various unit costs gives the gross unit total cost of producing gasoline (when all input prices are held constant at 1939 values). This gross unit total cost declined from $17.6204 per 100 gallons (or 17.62 cents per gallon) in the 1914 Burton thermal cracking process to $5.5441 per 100 gallons (or 5.54 cents per gallon) in the 1955 fluid catalytic cracking process. Measured gross, unit total cost (at constant 1939 prices) was reduced by just over two-thirds. This represents a reduction of 2.8 per cent per annum compound.

Note that the reduction in unit materials cost contributed just over one-half to the reduction in the gross unit total cost of production. This happened not only because of the substantial proportionate reduction in the unit input requirement for crude petroleum, but also because the cost of crude petroleum is the largest single item in total cost.

This illustrates the general principle that research and development aimed at reducing the unit input requirement for those resources which form the greater part of cost are likely to be more rewarding. In the case of petroleum refining, it was clear from the start that efforts to increase the efficiency of extracting gasoline from crude petroleum would be very rewarding, and much of the research effort was, accordingly, directed towards that aim. But as economists we must clearly understand why this was so, and the answer clearly lies in the proportionate cost structure (down the unit cost columns) shown in the top part of Table 5.3.

These figures on unit cost are 'gross' in the sense that they do not allow for the impact of the recovery, by sale, of by-products (middle distillate and fuel oil). Such recovery of by-products can be understood as reducing the materials input cost for the main product (gasoline). Consequently, the value of by-products should be deducted from the gross unit total cost to give a net unit total cost pertaining to the production of gasoline only.

In the bottom part of Table 5.3, figures are given on the recovery of middle distillates (kerosene) and fuel oil from the quantity of crude petroleum required to produce 100 gallons of gasoline. In quantity terms, each recovery was reduced, especially the residual fuel oil. This is consonant with a reduction in 'waste', or at least a reduction in the quantity of less remunerative products of the refining when a greater quantity of the more remunerative product is extracted.

Consequently, the dollar amounts (measured at constant 1939 prices) which should be netted off the gross unit total cost of gasoline decreased from \$7.0138 per 100 gallons in 1914 to \$2.2437 per 100 gallons in 1955 (all at 1939 prices). This decrease is the counterpart of more 'efficient' refining – 'efficient' in the sense of producing more of the high-value gasoline and less of the low-value by-products.

The net unit total cost of producing gasoline is shown in the last row of Table 5.3.

The net unit total cost of gasoline decreased from \$10.6066 per 100 gallons (or 10.61 cents per gallon) in 1914 to \$3.3004 per 100 gallons (or 3.30 cents per gallon) in 1955.

This is a reduction of just over two-thirds during the forty-one years, or a rate of decrease of 2.8 per cent per annum compound. The proportionate change in the net unit total cost is the same as the proportionate reduction in the gross unit total cost.

Substantial and sustained reduction in unit cost over such a long period demonstrates the power of technological change to reduce the unit total cost of production.

What did all this mean for the consumer? To answer this we can look at gasoline prices. This has the merit of getting away from estimates of unit cost to actual observations on prices ruling in the market.

Table 5.4 shows that the wholesale price of gasoline (excluding taxes) in the United States was 13 cents per gallon in 1914, declined to 9.58 cents per gallon in 1939, and then increased to 16.18 cents per gallon in 1955. The retail price (excluding taxes) was almost the same in 1955, 21.42 cents per gallon, as it had been nearly thirty years earlier in 1926.

These price comparisons are, of course, affected by the changing general value of money. If we adjust for general price inflation by deflating gasoline

Table 5.4
Gasoline prices in the United States, 1914 to 1955

Year	Gross national product implicit price index, 1958 = 100	Cents per gallon, current values, excluding tax, regular grade			Cents per gallon, constant 1958 values, excluding tax, regular grade		
		Price at refinery[(a)]	Wholesale dealers' net price[(b)]	Retail price[(c)]	Price at refinery	Wholesale dealers' net price	Retail price
1914	30.7	8.4[(d)]	13.00	n.a.	27.36	42.35	n.a.
1922	50.1	11.8[(e)]	22.63	n.a.	23.55	45.17	n.a.
1926	51.1	9.75 (57)	17.44	20.79 (55)	19.08	34.13	40.68
1939	43.2	4.13 (68)	9.58	13.31 (74.5)	9.56	22.18	30.81
1942	53.0	5.75 (73)	10.44	14.46 (78.0)	10.85	19.70	27.28
1955	90.9	10.50 (84)	16.18	21.42 (88.8)	11.55	17.80	23.56

(*a*) Midwestern refineries, markets as of first week in March each year. Figure in brackets is octane rating (motor) except for 1955, which is the research rating.
(*b*) For 1914 and 1922, average tank-wagon price; 1926 to 1955, average of fifty cities in the United States.
(*c*) Average service station price, exclusive of taxes. Figure in brackets is octane rating (research); first figure on octane rating is for 1925.
(*d*) Average refinery price.
(*e*) Average refinery price, 1923.

Sources: GNP implicit price index: United States Department of Commerce Bureau of the Census, *Historical Statistics of the United States Colonial Times to 1970*, Part 1, Series F 5, p. 224; gasoline prices, current values: John Enos, *Petroleum Progress and Profits: A History of Process Innovation* (The M.I.T. Press, 1962), tables 2a, 2b and 2c, pp. 287–91; 'n.a.' means 'not available'.

prices by the implicit price index for the gross national product of the United States, based on 1958 prices or the '1958 value of money', we see a pattern of almost steady decline in the 'real' price of gasoline; that is, in the price of gasoline relative to other prices generally in the economy.

The 'real' price of gasoline at the refinery decreased from 27.36 cents per gallon (1958 values) in 1914 to 11.55 cents per gallon (1958 values) in 1955. The 'real' wholesale price of gasoline decreased from 42.35 cents per gallon (1958 values) in 1914 to 17.80 cents per gallon (1958 values) in 1955. (The wholesale price is greater than the refinery price because of transport costs etc. from refinery to wholesaler.) Each of these represents a reduction of 2.1 per cent per annum compound.

The 'real' retail price of gasoline decreased from 40.68 cents per gallon (1958 values) in 1926 to 23.56 cents per gallon (1958 values) in 1955. (The retail price is greater than the wholesale price because of the distributor's margin and the retailer's margin.) This represents a reduction of 1.9 per cent per annum compound. (Note that all these prices exclude taxes, because taxes are not relevant to the economics of production and resource use.)

All these price decreases are proportionately very substantial, and we must conclude that the consumer has enjoyed a great benefit in the form of lower 'real' prices resulting from technological change.

Finally, we have to consider that these measures of the change in unit cost and in product price are inadequate indicators of the impact of technological change (and economies of scale). This is because the octane rating of gasoline was improved by nearly three-quarters. This octane increase permitted an improvement in the performance of motor vehicle engines.

Table 5.5 shows that the productivity of a gallon of gasoline, measured in terms of the ton-miles which could be carried by 1 gallon of gasoline at a constant speed of 40 miles per hour, increased from 24 ton-miles in 1914, to 29.2 ton-miles in the 1920s, to 49.5 ton-miles in 1955. That is, a gallon of 1914 55-octane gasoline produced by the Burton thermal cracking process would carry 1 ton for 24 miles at 40 miles per hour; a gallon of 1920s 73-octane gasoline produced by the continuous thermal cracking process would carry 1 ton for 29.2 miles at 40 miles per hour; and a gallon of 1955 95-octane gasoline produced by the fluid catalytic cracking process would carry 1 ton 49.5 miles at 40 miles per hour; in all instances given an engine designed to use gasoline of that octane rating. You can see how the octane rating was a matter of great importance to motorists.

It is clear that the motorist was getting more 'service' from a gallon of gasoline in 1955 than in 1914, and this increased quality per gallon is not

Table 5.5
Quality improvement in gasoline due to technological change, 1913 to 1955

Process and year	Gasoline octane rating[(a)]	Productivity of gasoline: ton-miles carried at 40 miles per hour by 1 gallon
Burton process; 1913[(b)], 1922	55	24
Continuous thermal cracking process; 1922, 1946	73	29.2
Houdry process; 1936, 1939–40	87	39.5
Fluid process; 1942, 1955	95	49.5

(*a*) Research rating; indicates 'quality' of gasoline.
(*b*) Date is for original 1913 Burton process; figures relating to 1914 used in Tables 5.2 and 5.3 refer to (improved) Burton process with false bottom plates; each process produced gasoline of 55 octane rating.

Source: John Enos, *Petroleum Progress and Profits: A History of Process Innovation* (The MIT Press, 1962), table A, p. 271.

taken into the reckoning when we look simply at unit cost per physical gallon or at price per physical gallon.

So how can we adequately assess the impact of technological change in petroleum refining? One way would be to look at the ton-miles carried per dollar of (constant price) cost incurred in producing gasoline. Another way would be to look at the ton-miles carried per dollar (at constant 1958 values) spent by the consumer on buying gasoline (retail). This is a ratio of the service or 'benefit' which is wanted from gasoline either to the 'real' cost (in 1939 prices) incurred by the refinery in producing the gasoline or to the 'real' price (in 1958 values) paid by the consumer.

Accordingly, from Tables 5.3, 5.4 and 5.5, we could do a little arithmetic as follows:

Analysis in terms of 'real' unit cost	*1914*	*1955*	*Change, per cent per annum compound*
Unit cost per 100 gallons of gasoline, $ at 1939 prices:			
Gross (not allowing for by-product recovery)	17.6204	5.5441	–2.8
Net (allowing for by-product recovery)	10.6066	3.3004	–2.8
Ton-miles carried for that cost	2,400	4,950	1.8
Unit cost of motoring (gasoline only), cents per ton-mile, 1939 prices:			
Gross	0.7342	0.1120	–4.5
Net	0.4419	0.0667	–4.5
Analysis in terms of 'real' retail price	*1926*	*1955*	
Retail price of gasoline, cents per gallon, 1958 values	40.68	23.56	–1.9
Ton-miles carried for that price	29.2	49.5	1.8
Unit price of motoring (gasoline only), cents per ton-mile, 1958 values	1.3932	0.4760	–3.6
'Gasoline price-productivity': ton-miles per consumer-price dollar, 1958 values	71.78	210.10	3.8

This shows that, over the period 1914 to 1955 and in terms of the 'real' cost of producing gasoline, technological change (combined with economies of scale) reduced the unit (ton-mile) cost of motoring (gasoline only) by 4.5 per cent per annum compound on both the gross and net cost.

Over the period 1926 to 1955 and in terms of the 'real' retail price (excluding taxes) paid by the motorist, the per unit (ton-mile) price of motoring (gasoline only) was reduced by 3.6 per cent per annum compound.

Conversely, the 'service' purchased by a consumer-dollar (at 1958 values and excluding taxes) when buying gasoline increased from 71.78 ton-miles in 1926 to 210.10 ton-miles in 1955, a rate of improvement of 3.8 per cent per annum compound. (This also depended, of course, on the interdependent and complementary improvement in automobile engines to take advantage of the improved quality gasoline.)

These latter rates of change are arguably the best estimates we can make of the impact of technological change as it affected the user of gasoline. But these rates of improvement depended upon, and can only be understood in the context of, the reduction in the unit cost of gasoline production, whether taken gross or net.

Thanks to the new-plant technological changes brought about by the work of innovators like Burton, Clark, Houdry, and the fluid catalytic cracking research team, and thanks to the in-plant technological changes brought about by the many others working in petroleum refining, the economy now gets substantially more ton-miles of motoring output per unit of resource use in petroleum refining, and the consumer enjoys a substantially improved standard of living in terms of motoring transport.

PART II

THE LEARNING CURVE

6 The Learning Effect

6.1 Learning in a plant

In this part we consider the learning effect, and how the learning effect may be analysed through the learning curve and the time constant progress function.

The learning effect is sharply distinguished from technological change because the latter involves physically different equipment in distinct or physically altered plants.

By contrast, the learning effect refers to what happens in one single, or 'given', plant with physically unchanged equipment as it proceeds with the production of a given type, or 'model', of product.

The learning effect occurs as a given worker or a given group of workers 'learns' how better to operate a given piece of equipment or a given set of pieces of equipment (a plant). Such learning manifests itself in improved performance in the same plant. At first, we shall use the word 'performance' rather generally – because there are various possible ways of measuring performance. Subsequently in this chapter and in the next chapter, we shall consider explicitly various specific measures of performance pertinent to learning.

The term 'learning effect' means the general sorts of things behind, and the consequences for labour productivity of, 'learning' in a given plant; the term 'learning curve' refers to a specific (mathematical and statistical) technique of, or function for, describing the learning effect in a given plant.

A given plant may be defined with respect to at least six things: (1) technology; (2) designed scale of output per annum; (3) fixed capital equipment and degree of capital-intensity (including also plant layout and the organization of production); (4) labour force (same personnel), and their methods of work and their skills; (5) materials used, including materials handling and inventory control; and (6) product, meaning type and design of product, or 'model'.

The learning effect concerns mainly, but not entirely, the methods of work and the skills of the same personnel. The learning effect may include some *minor* improvements to plant layout and the organization of production and/or some *minor* improvements to materials handling and/or inventory control, but neither of these changes must be such as to be capable of being considered a process innovation.

In regard to methods of work, it would be preferable to have a situation in which, at the outset of production, the best-practice motion and time study (or ergonomic 'work study') methods of work had been established in a plant which had workers all of whom were well trained. Then we could study a 'pure' learning effect which simply improved the performance of initially well-trained workers who had already started by working with best-practice work study methods.

Work study (or, synonymously, 'motion and time study') refers to all techniques involving a systematic study of work and working arrangements with a view to ascertaining the one best way of doing or accomplishing anything; to standardizing those arrangements; and to improving the training of workers.

However, it is often the case that even supposedly 'well trained' workers have to learn various aspects of work 'on the job', and that best-practice work methods have to be established by trial and error during the course of actual production. In such a case, we have to accept such improvements due to on the job training and/or to changes in methods of work as integral parts of the process of learning.

Figure 6.1 shows how we shall analyse the performance of a given plant. We start by supposing that in the stage before production begins in a plant, the best-practice methods of work have been arranged, the capital–labour ratio (or capital-intensity of production) has been determined, the technology is the most advanced possible, and the scale of production has been established. We assume that capital-intensity, technology, and scale of output all remain unaltered, not because this is a realistic assumption but because we wish to concentrate upon the impact of learning.

Logically, the learning effect can start only when, or after, the plant begins production. The performance of workers improves because of experience gained through actual production. Furthermore, managers may improve the arrangements for, or organization of, work through work study or other methods. But, in principle, the learning effect is distinct from such other improvements; that is, either can occur without the other, even though they often occur in association.

There are then two methods by which we may describe and analyse the performance of a plant. The first method is to relate performance to the passage of elapsed time. The second method is to relate performance to the cumulative total of output (ever) produced by the plant.

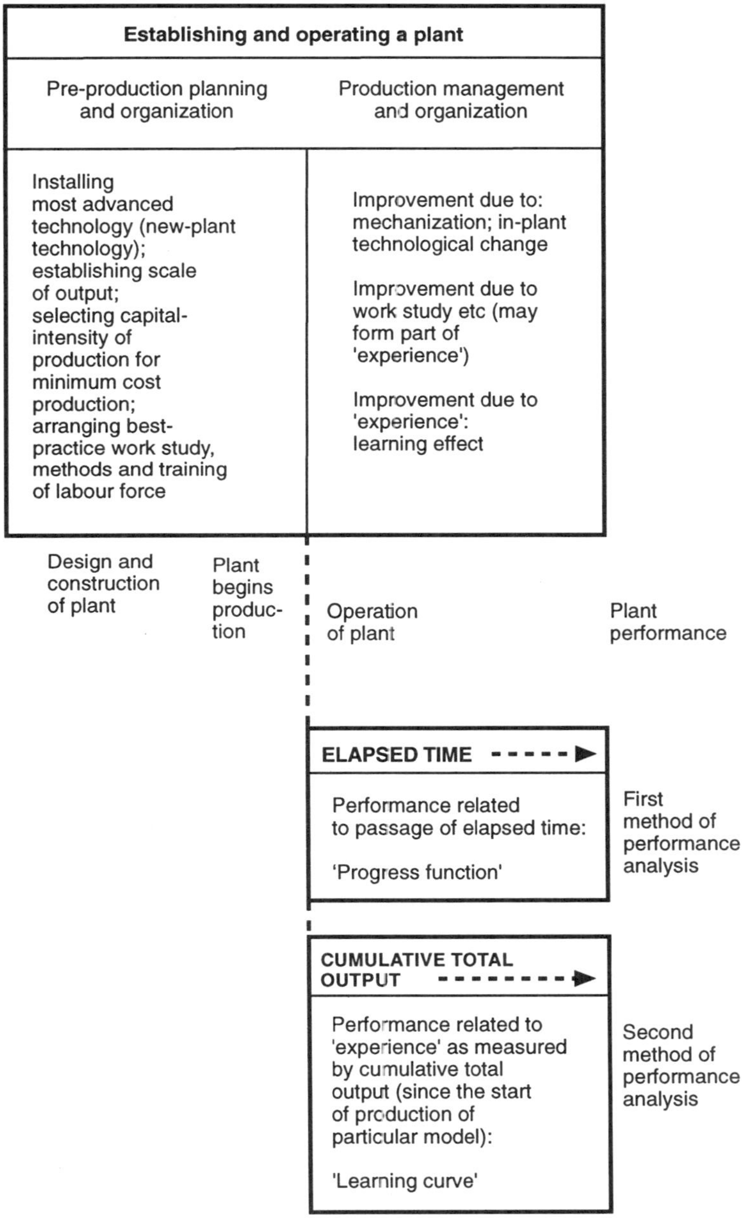

Figure 6.1 The progress function and the learning curve as two methods of performance analysis

We digress to note that we must now be explicit and careful with regard to the word 'time'. On the one hand, we have elapsed time, which is time elapsing on a clock (or on a calendar if a long period of elapsed time is being considered). On the other hand, we have labour-time, where labour-time may be measured as the number of labour-hours used in the process of production.

The distinction between elapsed time and labour-time is straightforward and obvious. The input of two workers working simultaneously and uninterruptedly for *one* hour of elapsed time constitutes an input of *two* hours of labour-time.

Only in the case of a single worker working alone and uninterruptedly do elapsed time and labour-time coincide. (I make these cautionary remarks about 'time' because the practice in work study, or time and motion study, of dealing with only one worker leads to careless and confusing use of the word 'time' where no explicit distinction is made between elapsed time and labour-time.)

If the performance of a plant is related to periods of elapsed time, then I call this method of performance analysis a 'progress function'.

Relating the performance of a plant to periods of elapsed time is helpful for various purposes (as we shall see), but it does not help us much in *explaining why* performance has changed. One cannot really say that the passage of elapsed time has *caused* the change in performance. If performance in a given plant has changed from one period to the next, the immediate question is: Why?

This question is answered by the second method of performance analysis – by the learning curve.

In the present context, the word 'learning' means a (relatively) permanent change in skills and/or knowledge and/or abilities, which change results mainly from practice or experience and partly from observation, training, or education of various sorts, and the results of which change we can observe in the measured statistics relating to the process of production.

In the present context, 'skill' means a learned ability or behaviour pattern which, being learned, is in some sense the opposite of an instinctive or innate ability or behaviour pattern. Skill also involves processing information and working purposefully towards some understood objective. In the context of work, the development of skills is one of the most important components, albeit not the only component, of learning.

If the performance of a given plant improves from one period to the next, then the answer to the question 'Why?' may be found in learning.

Learning causes the improvement and, in this context, learning itself results from practice and experience.

In this context, the words 'practice' and 'experience' have different but related meanings. The word 'practice' means repeating an operation, generally with a view to improving coordination and/or speed. The word 'experience' signifies that one has had the requisite practice and also that one has put that practice into effect, generally more than once. Thus a football team gains practice in various ways, but has experience of playing football matches.

In the process of production, practice and experience is gained in each performance of the cycle of tasks required for production. It is therefore appropriate to measure practice and experience *cumulatively* in terms of the total number of times (cumulatively) which each cycle of operations has been performed (since the start of production of that particular model of product in the plant).

In this way, we acknowledge that practice and experience (taken as a unified combination) is essentially a cumulative phenomenon, in which each cycle or 'bit' of practice and experience reinforces all previous cycles or 'bits' of practice and experience. As the saying goes: 'Practice makes perfect'.

One way to measure the quantity of cycles or 'bits' of practice and experience is to take the *cumulative total* of output ever produced (of that particular model) by the given plant.

If we relate performance in this way to cumulative total output, then we are using what should properly be called a 'learning curve'. Because there are different ways of measuring performance (as we shall see), there are accordingly different forms of the learning curve, each depending on the particular measure of performance.

(The term 'learning curve' in the application known today appears to have originated in, or at least to have been popularized by, a 1919 psychological study, relating to students on a type-writing course, of improvement in typing speeds as a function of practice in typing. (Earlier precursors of mathematical learning functions are discussed in E.R. Hilgard, *Theories of Learning* (Methuen, 1958), Chapter 11, 'The Emergence of Mathematical Models'.) The 1919 study was that by L.L. Thurstone, *The Learning Curve Equation, Psychological Monographs*, vol. XXVI, no. 3, 1919. In this study, for each student 'the learning curve . . . is plotted between attainment . . . [measured] in terms of the number of words [type-] written in a four minute test given weekly for seven months, and formal practice . . . [measured] in terms of the total number of pages [type-] written since entering the course' (p. 13). The derivation of the particular forms of learning curve with which we shall be concerned and the application of the learning curve to the analysis of manufacturing production are due to T.P. Wright, 'Factors Affecting the Cost of Airplanes', *Journal of the Aeronautical Sciences*, vol. 3, no. 4, February 1936, but Wright's 1936 paper did not use the specific term 'learning curve'; instead Wright spoke more specifically of a 'curve which shows [the] variation of [unit] labor cost with production quantity' (p. 124).)

Obviously enough, learning affects the individual worker. But learning is also capable of affecting a group of workers, or a plant as a whole. These issues will now be considered in turn.

6.2 Learning by an individual worker

Table 6.1 shows an example of the learning effect as it applies to an individual worker.

In this example, the operation, or 'cycle of tasks', was assembling four parts of a mechanical pencil (what used to be known as a 'propelling pencil' of the old-fashioned kind). The only special equipment used was a steel plate containing a V-block to hold one of the parts while the mechanism was being assembled. Throughout, the observations relate to one individual worker.

The work was arranged according to the best work study practice, with four boxes, each containing the separate parts, placed conveniently to hand, and another conveniently placed box to receive the assembled mechanisms. Ralph M. Barnes, *Motion and Time Study Design and Measurement of Work* (Wiley, Seventh Edition, 1980) has, on p. 546, a photograph showing the worker at work and the arrangements of the workstation (fig. 298) and a diagram showing the mechanism to be assembled (fig. 297). These work arrangements were not subsequently changed, so Table 6.1 presents a pure learning effect.

Note the convention of numbering the weeks (periods) from zero, rather than from 1. This has the advantage that the number then also tells you how many complete weeks (periods) of complete production have already occurred. For example, numbering the *second* week of production '1' tells you that 1 complete week of production has already occurred. In the chapter dealing with the time constant progress function we shall see that this method of numbering is mathematically required.

During the first week (Week 0), 9,300 pencils (more accurately, the internal mechanisms of propelling pencils) were assembled by this worker, who worked 40 labour-hours during the week. During the second week (Week 1), 11,900 pencils were assembled, also during a working week of 40 labour-hours. And so on. You can see that the number of mechanisms assembled per week

Table 6.1
Learning effect for a worker assembling a mechanical or 'propelling' pencil mechanism[(a)]

Week, t	Labour-hours worked per week, L_t	Number of mechanisms assembled during that week, Q_t	Unit labour requirement: labour-hours per thousand mechanisms assembled,[(b)] (L_t/Q_t) 1,000	Cumulative total number of mechanisms assembled: 'practice' or 'experience'[(c)]
0	40	9,300	4.30	9,300
1	40	11,900	3.36	21,200
2	40	14,600	2.74	35,800
3	40	15,300	2.61	51,100
4	40	15,800	2.53	66,900
5	40	15,900	2.52	82,800
6	40	16,200	2.47	99,000
7	40	16,000	2.50	115,000
8	40	16,300	2.45	131,300
9	40	16,400	2.44	147,700
10	40	16,000	2.50	163,700
11	40	17,000	2.35	180,700

(*a*) Worker assembles four parts (spiral, lead carrier, bushing, and stem) as internal part of mechanical (or 'propelling') pencil. Note that the data relate to the same operator throughout the twelve weeks.

(*b*) Average for given week. Note very carefully, for future reference and as a standard for comparison, this by-now familiar measure of unit labour requirement. To be explicit, it is a measure of unit labour requirement which is averaged over whatever quantity of output happens to have been achieved during the particular period (the period in this case being a week). We could therefore explicitly call this a 'period average unit labour requirement'. So far in this book we have dealt only with period average unit labour requirement or its converse, period average labour productivity. However, we shall shortly have to deal with, and understand, other and different measures of unit labour requirement.

(*c*) By the end of the week, starting from the beginning of the first week. This column gives a sample of values of i (namely, those values of i reached by the end of that particular week), where i refers to the serial number, in sequence, of the item produced.

Source: Ralph M. Barnes, *Motion and Time Study Design and Measurement of Work* (Wiley, Seventh Edition, 1980), pp. 542–8, 561–6, Fig. 299 (quoting a study by L.F. Youde, 'A Study of the Training Time for Two Repetitive Operations', University of Iowa Thesis, 1947).

increased to 17,000 mechanisms per week.

In other words, performance improved. This is the learning effect: in this case, improved performance by an individual of his or her work.

It is important to note that the improvement shown in Table 6.1 must be attributed to 'learning', to an improvement in the skill and speed of the worker. There was no mechanization, no technological change, no modification to the product, no change in plant layout, no work study improvement. In this example there is nothing other than learning which could explain the improvement in this individual worker's performance.

It is important also to note the magnitude of the improvement – by over three-quarters. The learning effect is not confined to merely quantitatively minor improvements in productivity: it is capable of effecting quantitatively major improvements.

It is furthermore important to note that Table 6.1 does not represent an isolated and rare occurrence. This sort of thing happens all the time.

After the twelfth week – data not shown in Table 6.1 – performance reached a 'plateau'; that is, the performance remained more or less constant. Such plateaus are important, but it is not necessary in the present context to study this specialist phenomenon, partly because the purpose of Table 6.1 is to illustrate the initial phase of the learning effect and partly because the plateau phenomenon is in any case largely incorporated in the form of the learning curve which we shall subsequently derive.

We should consider the two measures of performance presented in the second and third columns of Table 6.1 (the weeks are *row* headings, not data).

The first measure of performance is labour productivity. This is the number of mechanisms assembled per worker per week. To be very specific, this is a measure of *period average* productivity. That is, labour productivity is measured with respect to the average productivity during some specified period of elapsed time, in this case a week.

The second measure of performance is unit labour requirement. This is the number of labour-hours required, on average during the period, per unit of output. This *period average* (in this case, week average) unit labour requirement is calculated by dividing the total number of labour-hours worked during the period (a week) by the total quantity of output produced during the period (a week).

(The reason for now being explicit about the 'period average' nature of the measure of performance is that we shall soon be working with averages calculated with respect not to periods of elapsed time, but with respect to batches of output. These we shall call 'batch averages' and such batch averages will be fully explained in due course.)

The data of Table 6.1 can be analysed through either of two methods of performance analysis.

In the first place, we can relate performance to elapsed time. We may do this, as shown in Figure 6.2, by charting the improved performance week by week. To do this we can use either measure of performance: period (week) average labour productivity, in the top part of Figure 6.2, or period (week) average unit labour requirement, in the bottom part of Figure 6.2. Which measure we use is largely a matter of convenience to suit whatever purpose we may have, but we shall subsequently concentrate upon unit labour requirement because this measure is best suited to an analysis of unit cost.

Either part of Figure 6.2 represents what I call a 'progress function'; that is, a function which relates performance to the passage of elapsed time.

Looking at the progress functions in Figure 6.2, we can readily see that at first performance improves rapidly, but that the rate of improvement tends to slow down with each passing period (week). This is a typical pattern.

The presentation of the progress functions in Figure 6.2 is relatively straightforward. Average performance during each week is related to that week. The matching of coordinates in the graph is not problematic; we have the following matches for labour productivity (units of measurement taken as read): Week 0, 9,300; Week 1, 11,900; Week 2, 14,600; and so on. Alternatively, we have for unit labour requirement: Week 0, 4.30; Week 1, 3.36; Week 2, 2.74; and so on.

But these progress functions simply describe what is happening with regard to the passage of time, or over successive periods. Progress functions are useful to this extent, but they do not help us much with answering the question: Why did performance improve?

If we wish to answer this question, we need to relate performance to the worker's practice or experience which has caused the learning, the impact of which learning may be measured either by the increase in period average productivity or by the reduction in period average unit labour requirement.

We cannot, nor should we attempt to, measure practice and experience simply by the passage of elapsed time for the following reason. Practice and experience is gained in each performance of the cycle of tasks or operations; in this case, in each mechanism assembled. Thus, we could say that the worker had gained 9,300 'bits' of practice and experience during the first week and 11,900 'bits' of practice and experience during the second week. Consequently, the first week and the second week as equivalent slices of elapsed time do not represent equivalent total 'bits' of practice and experience; and consequently it is not correct to think that a week, as marked on the horizontal axis, represents practice and experience.

If practice and experience is to be measured by each cycle of operations,

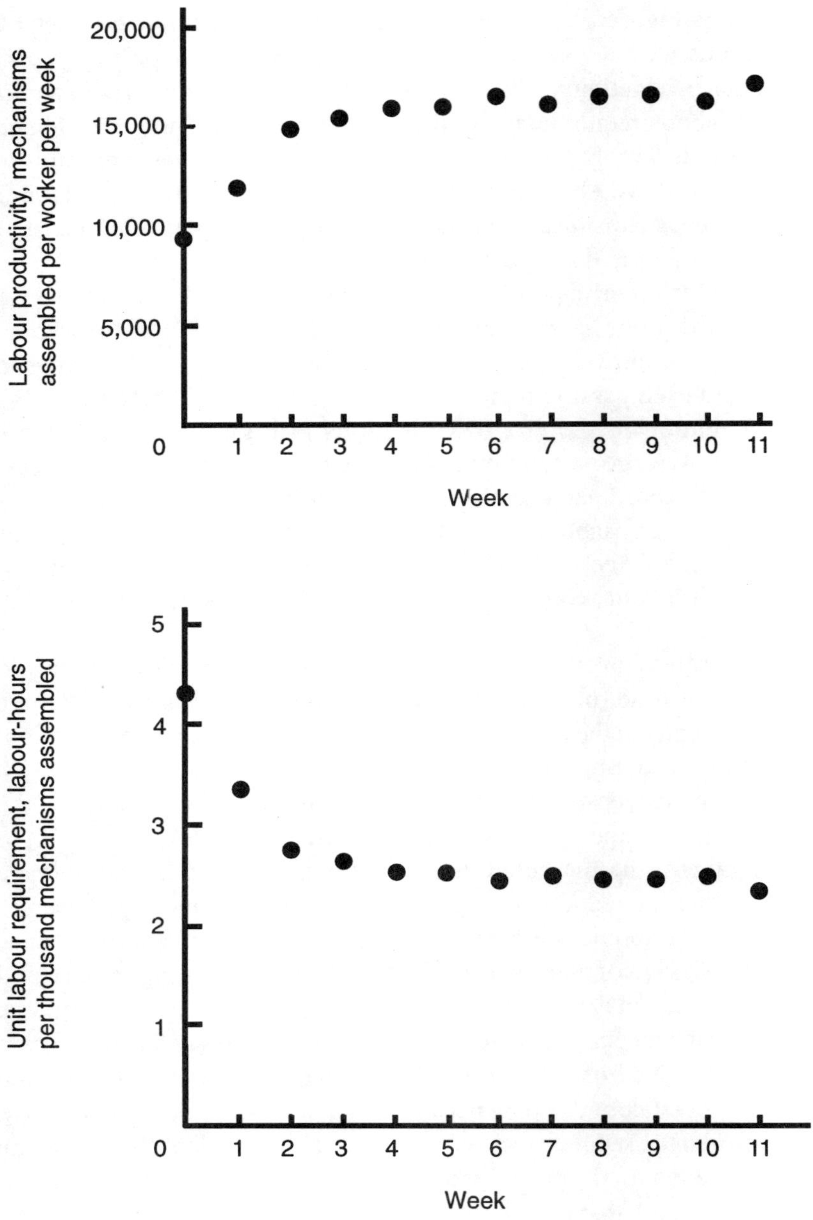

Figure 6.2 Progress functions for pencil mechanism assembly

and if the effect of practice and experience is cumulative, then what it is appropriate to say about this worker is the following.

By the end of the first week, this worker has had 9,300 cycles or 'bits' of practice and experience of assembling mechanisms. By the end of the second week, this worker has had 21,200 (= 9,300 + 11,900) cycles or 'bits' of practice and experience. By the end of the third week, this worker has had 35,800 (= 21,200 + 14,600) cycles or 'bits' of practice and experience. And so on.

It is thus appropriate to measure practice and experience *cumulatively* in terms of the total number of times (cumulatively) which each cycle of operations has been performed (since the start of operations). In this way, we acknowledge the truth that practice and experience is essentially a cumulative phenomenon, with each cycle or 'bit' of practice and experience reinforcing *all* previous cycles or 'bits' of practice and experience. Furthermore, working with cumulative total output deals with the problem that equivalent slices of elapsed time may contain different amounts of practice and experience.

In Thurstone's 1919 learning curve study of students learning to type, practice was measured *cumulatively* in terms of the total number of pages type-written since entering the typewriting course.

Accordingly, we shall measure practice and experience by *cumulative total output*. The last column of Table 6.1 gives the cumulative total number of mechanisms assembled since the start of operations to the end of that week. This column is derived simply by adding, cumulatively, down the column of weekly assemblies.

We may now try to present the data by relating performance to the cumulative total of output produced. This is the second method of performance analysis.

However, if we try to do this with the period (week) average data in Table 6.1 we run against a considerable difficulty. It is important to understand this difficulty, because the solution to the difficulty results in the proper derivation of the learning curve. Sometimes you will see in the literature an attempt to present or display the learning curve in terms of period averages, such as those given in Table 6.1, but this leads to considerable confusion and is symptomatic of a failure fully to understand the learning curve.

The difficulty or problem can be explained as follows.

Suppose that, as shown in Figure 6.3, we mark off cumulative total mechanisms assembled along the horizontal axis and then mark off, as in Figure 6.2, performance along the vertical axis. Suppose that we then plot period (week) average labour productivity or period (week) average unit labour requirement against cumulative total number of mechanisms assembled (top

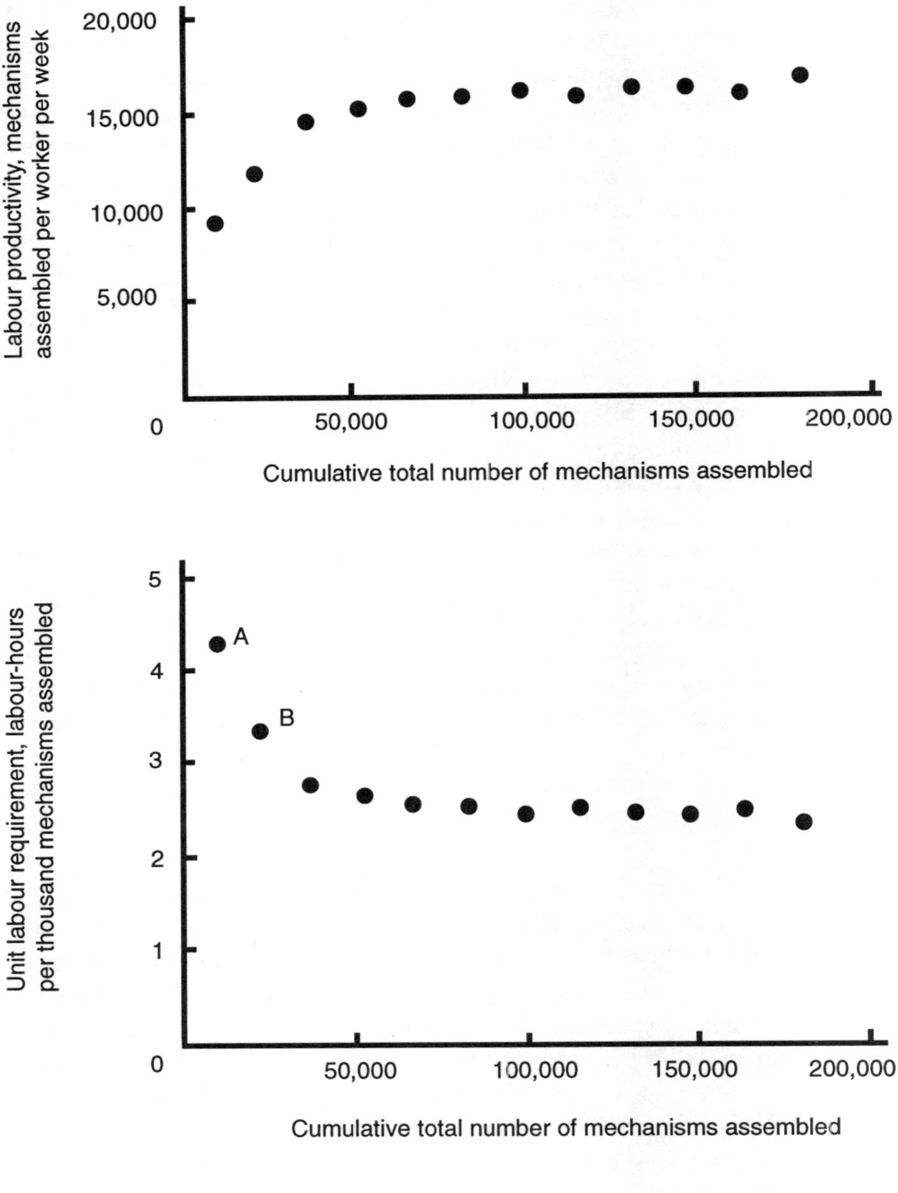

Figure 6.3 Performance and experience in pencil mechanism assembly

and bottom parts of Figure 6.3 respectively). We then obtain the pattern of coordinates as marked.

This seems to show a pattern of performance improving rapidly at first and then slowing down. But we would be seriously in error if we were to accept this presentation or display as correct. Neither part of Figure 6.3 represents a learning curve; in each part of Figure 6.3 the matching of coordinates is highly problematic. Why?

The answer is that, at each cumulative total (along the horizontal axis), the vertical coordinate is an average for the recent past, in this case the past week. Now, and this is the first step of the argument, the period of duration of the recent past is arbitrarily imposed by the way in which the data happen to have been recorded and reported (week by week).

If, and this is the second step of the argument, the data had been recorded day by day, and we thus had day average productivity or day average unit labour requirement, together with cumulative total mechanisms assembled by the end of each day, then plotting performance so measured against cumulative total mechanisms assembled would give us an entirely different curve.

To illustrate this, consider points A and B in the bottom part of Figure 6.3. Coordinates such as these are often (but erroneously) displayed as part of what is (erroneously) called a 'learning curve'.

Figure 6.4 repeats, on an enlarged scale, this part of Figure 6.3, where the week average coordinates at A and B are plotted as before but with a horizontal line to indicate the level of the period average. We then present the ten coordinates C to L as these coordinates would appear had a daily record of output been available so that day average unit labour requirement could have been calculated for each of the five days in the first and second week, together with the cumulative total mechanisms assembled by the end of each day. (These coordinates come from Table 8.3, to be presented in the last chapter, and so their derivation will be fully explained in due course; for the moment you can accept these coordinates as well established according to the (properly derived) learning curve.) Note that the *horizontal* coordinates at A and G (after 5 days) and at B and L (after 10 days) do not match exactly because A and B are empirical (actual) observations whereas G and L are estimated (theoretical) values. However, G is only 1.8 per cent horizontally different from A, and L is only 2.4 per cent horizontally different from B.

You can see that the curve which could be traced through coordinates C to L inclusive would be in an entirely different position from the curve which could be traced through coordinates A and B. Consequently, each such curve

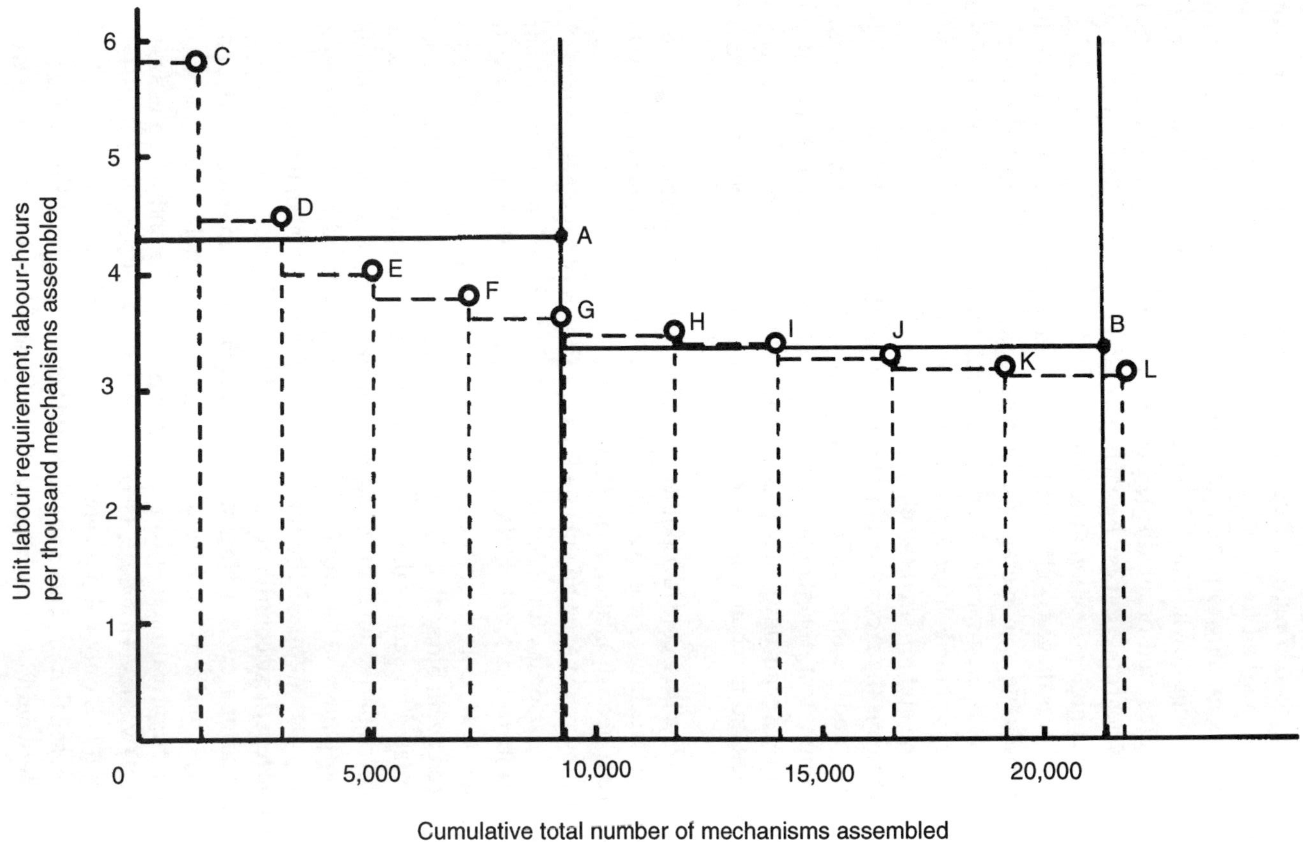

Figure 6.4 The arbitrary position of period averages against experience

is an arbitrary curve, meaning that its position depends upon the periods for which the observations have been collected and reported. As a corollary, neither curve can be a learning curve, because the position of the learning curve should not vary arbitrarily in this way.

The only way in which we can escape from this arbitrariness is to have a measure of performance item by item. In this case we could relate performance achieved during the very last item produced to the cumulative total number of items produced. For example, we could then relate the unit labour requirement for the 9,300th mechanism assembled to 9,300 mechanisms cumulatively assembled, and so on. Only when we get down to working on an item-by-item basis do we get away from an averaging of performance during an arbitrarily selected recently past period.

However, given the way in which production data are generally recorded and reported we are unlikely ever to have observations on an item-by-item basis. Despite this, all is not lost, because there is an analytic technique which will enable us to use the data of Table 6.1 and at the same time exactly to derive, and properly to present, the learning curve. The difficulty described in Figure 6.4 is a difficulty which we will fully resolve in the next chapter.

For the moment we will turn to the matter of learning in a work group and so extend our discussion of the learning effect to an entire plant, or to parts thereof.

6.3 Learning by a work group

Workers generally work not alone but in work groups. As with an individual worker, the performance of a work group may improve with practice and experience.

A work group may be defined as two or more workers whose jobs are related and who work regularly in collaboration on part, or the whole, of the work process and who have developed certain recognized and acknowledged, stable and predictable, types of interaction with other members of the work group. Generally, the members of a work group regard themselves as constituting a work group.

The work group should be regarded as a system, the structure of which is determined by the actions of members. This is in preference to regarding a work group as a set of individuals (or a set of individuals occupying certain jobs), even though the members of a group tend, in a less abstract and less analytic way, to perceive and describe the group in terms of the individuals involved. (A 'system' approach means that one individual – say, a maintenance worker – can without confusion be regarded as belonging to more than one work group.) The work group may or may not be formally designated and recognized by the enterprise.

A framework useful for understanding the performance of a work group may be derived from considering the hypothetical potential maximum performance of the work group against the actual achieved performance of the work group. The difference between potential maximum performance and actual achieved performance may be called the 'process losses'.

These concepts may be illustrated by data from a famous rope-pulling experiment conducted in the 1920s. In this rope-pulling experiment, a rope over a pulley was pulled, in a horizontal direction, to raise a weight from the floor. The rope was pulled variously by one person alone, by a group of two, by a group of three, and by a group of eight, and the maximum weight which could be lifted was recorded. Table 6.2 gives the results of the rope-pulling experiment as reported in a German work study textbook of 1935.

One person pulling the rope alone could lift a weight of 63 kilograms; two persons pulling ('working') together could lift a weight of 118 kg; three persons pulling together could lift a weight of 160 kg; and eight persons pulling together could lift a weight of 248 kg. These weights represent the actual achieved performance of the person or group.

Let us assume that the individuals taking part in this experiment were all reasonably comparable and similar with regard to strength and fitness (an assumption which may be excused because we are using the data only for the purpose of illustration). Then it is the case that two persons pulling together should be able to lift double the weight which one person pulling alone could lift; three persons, treble; and eight persons, eightfold. In other words, the average weight lifted per person should be constant. But, as we can see from row (3) in Table 6.2, this is far from being the case.

We may now use these data to illustrate hypothetical potential maximum performance, actual achieved performance, and hence 'process losses'. In this experiment, and for each group, potential maximum performance can be computed as 63 kg (the weight lifted by one person pulling alone) multiplied by the number of persons in the group. This is given in row (4) of the table. Next, we can calculate process losses, measured in terms of kilograms, by subtracting actual achieved performance from potential maximum performance. This is given in row (5) of the table. Finally, we may show the extent of the process losses relative to potential maximum performance in row (6) of the table. We see that this relative loss increases to a substantial

Table 6.2
The rope-pulling experiment, illustrating process losses in group work

Actual data					
(1)	Number of persons in group	1	2	3	8
(2)	Total weight lifted, kg	63	118	160	248
(3)	Average weight lifted per person, (2)/(1), kg	63	59	53	31
Derived data					
(4)	Potential (maximum) productivity, 63 kg × (1), kg	63	126	189	504
(5)	Process losses, (4) – (2), kg	0	8	29	256
(6)	Process losses as percentage of potential (maximum) productivity, ((5)/(4))100, per cent	0	6	15	51

Source: Walther Moede, *Arbeitstechnik Die Arbeitskraft Schutz-Erhaltung-Steigerung* (Stuttgart, 1935), Figure 100, p. 189; see also J. F. Dashiell, 'Experimental Studies of the Influence of Social Situations on the Behavior of Individual Human Adults', in C. Murchison (Editor), *A Handbook of Social Psychology*, The International University Series in Psychology (Clark University Press and Oxford University Press, 1935) and I.D. Steiner, *Group Process and Productivity* (Academic Press, 1972), p. 81.

percentage, about one-half, as the size of the group increases.

These process losses, and the increasing relative extent of the process losses, may be ascribed either to increased difficulties of coordination (getting everybody to pull at maximum effort at exactly the same instant in exactly the same direction) and/or to the possibility that, in a group, each individual feels that a (slight) slackening on his/her part will not noticeably affect the outcome, so that the larger the group the greater the slackening of individual effort.

Process losses may generally be ascribed to insufficient experience and/or practice or inadequately developed expertise and lack of familiarity with processes. There may be a 'managerial' dimension to process losses in that management may not, initially, arrange for the fastest flow of work, or may not manage inventories properly, leading to interruptions to work. Equipment breakdown may also be due to improper use of equipment arising out of inexperience, and this too will lead to process losses. The causes of process losses are many, but in the present context we concentrate simply upon what may generally be identified under the heading 'lack of experience'.

Whatever the cause(s), the rope-pulling experiment illustrates both the concept of, and the relatively large possible extent of, process losses. We should not push this laboratory experiment, involving a simple collaborative task, too far in its application to work groups and to the outcome(s) of their work in the real world, but the experiment does clearly illustrate the concepts involved. In the real world, potential maximum performance would be assessed in a variety of ways (such as work study, standard costs, or comparisons with other plants – so-called 'bench-marking'). But once potential maximum performance is established, the rest follows.

Unfortunately, one thing was missing from the rope-pulling experiment. This is the impact of learning. There was no report of giving the individuals or groups the opportunity to 'learn'; in this context, to develop the muscular strength of individuals and to develop the coordination among individuals in groups through regular and sustained practice. But it is reasonable to speculate that had this been done and the results reported we would probably have seen two things.

First, the capacity of an individual pulling by himself/herself to lift a weight would have increased with 'exercise' (practice). This would be due to an increase in muscle strength and fitness, etc. The corollary of this is that there would have been a corresponding increase in the potential maximum performance of any group.

Second, the effort and coordination among individuals in a group (given that each group always comprised the same individuals) would have improved as the group became a more practised and experienced team. This would be due to such things as learning how to pull together in exactly the same direction at exactly the same instant, perhaps assisted by the development of 'leadership' in the group, and by the development of 'team spirit' which motivates each individual to give of his/her best. The improved effort and coordination would be likely to reduce the relative extent of process losses. Such speculation at least illustrates the possibilities of the learning effect as it would operate in this context.

We can transfer these concepts to a framework for analysing performance by a work group in a workplace. Figure 6.5 schematically depicts the framework in relation to a work group, where we may measure performance by productivity (or production – in this context it is not necessary to be too precise about the measure of performance adopted).

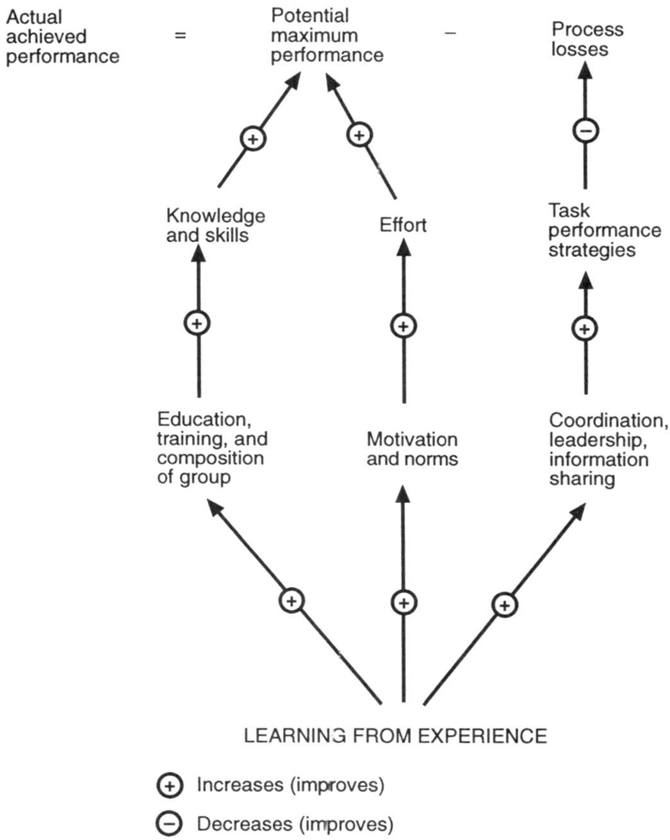

Figure 6.5 Group performance improvement

We defined and illustrated process losses as the difference between potential maximum performance and actual achieved performance. Figure 6.5 rearranges this equation to give actual achieved performance as the difference between potential maximum performance and process losses.

A work group has a potential maximum performance dependent upon its given resources (technology, scale of production, capital equipment) and the 'rules' under which it operates (for example, the length of the working week). But the work group's actual achieved performance may not equal its potential maximum performance because of what I call 'process losses'.

Process comprises all actions taken by individual members of the group or by the group itself when working with their resources to produce output. The neutral term 'process losses' can then be used to refer to any difference between potential maximum performance (however established) and actual achieved performance. (For a lengthier discussion of these issues, see Ivan D. Steiner, *Group Process and Productivity*, Academic Press, 1972.)

Taking as given the technology, scale of output, and capital-intensity of production, together with the 'rules' for production (because these are not matters which we wish to discuss in the present context), potential maximum performance may be seen as being determined (in part) by effort, itself dependent upon motivation and norms of the work group, and by knowledge and skills, each dependent in turn upon the composition and training of the work group.

Process losses may be seen as being determined (in part) by such things as inexperience, insufficient knowledge of processes, lack of coordination, undeveloped leadership, poor motivation, and inadequate information sharing among the group.

In turn, all these things may be each and severally improved by learning from experience.

Furthermore, such improvements can be sustained over a long period of time, leading in turn to sustained improvement in actual achieved performance over a long period of time. These are not matters for just the first few periods of production (although they may have their greatest impact during the first few periods of production).

When the impact of the sustained improvement (increase) in potential maximum performance is combined with the impact of the sustained improvement (decrease) in process losses, the result is necessarily a substantial, as well as a sustained, increase in actual performance (productivity).

We need to bear all this in mind as we turn to look at productivity in a given plant, a shipyard (Yard F), because we shall see quite astonishing increases in labour productivity.

6.4 The learning effect in the United States shipbuilding industry

The purpose of the preceding sections was to explain and illustrate some of the causes of the learning effect, both in relation to a single worker and also in relation to a group of workers, so that we could understand how and why the learning effect can have a substantial and sustained impact in a given plant.

We now look in detail at some important data which show the learning effect in given plants very clearly. The data relate to shipbuilding in the United

States during the Second World War. In the following chapter, we shall use these data to explain and illustrate the proper derivation of the learning curve.

The data are taken from the United States Department of Labor Bureau of Labor Statistics, *Wartime Employment, Production, and Conditions of Work in Shipyards*, Bulletin No. 824, 1945, hereafter cited as Bulletin 824. Some supplementary information is taken from relevant articles in the Department of Labor's *Monthly Labor Review* (November 1943, May 1944, December 1945, and June 1946), and from the standard official histories by Frederic C. Lane, *Ships For Victory A History of Shipbuilding under the U.S. Maritime Commission in World War II* (The Johns Hopkins Press, 1951), hereafter cited as *Ships For Victory*; and by Samuel Eliot Morison, *History of United States Naval Operations in World War II*, Volume I, *The Battle of the Atlantic September 1939–May 1943* (Little, Brown and Company, 1947). We may begin with the following observations:

> When Hitler met with his admirals in September 1942 to survey the submarine war, he confidently asserted that American shipyards could not build ships faster than they were being sunk. The Presidential directive of 24 million tons [ordered by Roosevelt to be constructed] for 1942 and 1943 was dismissed as mere propaganda. In fact, the race between construction and sinkings in 1942 was nip and tuck, but the submarines were defeated whereas the shipbuilders achieved their goal. They not only fulfilled the Presidential directive of 24 million, they built 27 million tons in 1942 and 1943. (*Ships For Victory*, p. 202.)

Behind this vital achievement in shipbuilding lies the story of the way in which the shipbuilders of the United States, substantially aided by the learning effect, responded to the demands made on them by the United States Maritime Commission and so provided (in President Roosevelt's words) the 'bridge of ships' across the Atlantic (and later across the Pacific) essential to the Allied war effort.

The United States Maritime Commission had been established in 1936 as an agency of Congress to assist the United States shipping industry, and by 1938 the Maritime Commission was actively engaged on its own account in commissioning vessels to be built to modernize the ageing US merchant fleet. Fortuitously and fortunately, this established a base of technical expertise in shipbuilding within the Maritime Commission.

With the outbreak of the Second World War in Europe, the demand for shipping increased and in response the Maritime Commission increased its orders for cargo vessels.

During 1940 the British merchant fleet was seriously depleted by submarine warfare (and to a lesser extent by aircraft attack, by surface ship attack, and by mines), and the British government requested the American government, as a matter of extreme urgency, to accelerate the programme of shipbuilding.

After Pearl Harbour (7 December 1941), this emergency programme of shipbuilding was very greatly increased as German submarines successfully attacked and sunk merchant vessels in the coastal shipping lanes between the mouth of the St. Lawrence and the Gulf of Mexico. In the first half of 1942 the loss in merchant vessels amounted to 6.4 million deadweight tons. (Deadweight tons is the weight in tons (2,240 pounds avoirdupois) of cargo which a vessel can carry without overloading.) To put this figure in perspective, in 1939 the entire merchant fleet of the United States totalled 12 million tons, and that of the world 81 million tons.

The size of the resulting programme of shipbuilding in the United States can be judged from the fact that the number of employees working in shipyards with contracts with the Maritime Commission increased from 47,300 in January 1941 to 650,900 at the peak in July 1943. At peak production, there were 267 building berths in shipyards working for the Maritime Commission as contrasted with 51 building berths in January 1941.

The main type of ship which was to be built under this emergency programme was officially designated as the EC-2, for 'emergency cargo vessel type 2', but became known as the Liberty ship. It is with this ship and this programme that we are concerned. By the end of the war, 2,708 Liberty ships had been built; the first one, named the *Patrick Henry*, being launched from the Bethlehem-Fairfield Yard, Baltimore, on 27 September 1941 and the last one being delivered on 30 October 1945.

The Liberty ship was 441 feet 6 inches (134.6 metres) in overall length and had a deadweight tonnage of 10,419 tons. In winning the race against submarine warfare it proved its value and became known as the 'workhorse of the fleet'.

The fact that a very large number of standardized ships were built in many different shipyards makes the Liberty ships programme an ideal subject of study for the learning effect.

The Maritime Commission had labour-hour cost contracts with the shipyards building Liberty ships. (In the wartime situation, the Maritime Commission was responsible for supplying the steel plates and other materials for construction to the shipyards, who thus became contractors for assembly using their own skilled labour and capital equipment.) Because of these labour-hour contracts, the Maritime Commission had records of the labour-hours worked in each shipyard. These data were subsequently collected together and published, shipyard by shipyard, in Bulletin 824.

It was noted that all the shipyards 'were able to make drastic cuts in the [unit] man-hours required as more vessels were completed' (Bulletin 824, p. 39). Furthermore:

> . . . the declines in [unit] labor requirements within individual yards with continued construction of standardized vessels have been remarkably steady . . . The general downward trend of average man-hour requirements . . . is due primarily to improvements in yard efficiency achieved as management and labor gained experience in constructing successive vessels of standardized types. . . . The experience gained by any shipyard, and hence the reduction in [unit] labor requirements, depends primarily on the number of vessels constructed, and not on the number of months the yard has been engaged in construction. (*Monthly Labor Review*, December 1945, pp. 1143–4.)

The official historian also recorded the opinion that the shipyards' success:

> . . . depended, first of all, at least from an engineering point of view, on having standardized the product. Speed in production came from building the same design over and over again with a continuity that made it possible both to learn from experience and to plan ahead. . . . on the average, man-hours per [Liberty] ship were cut almost in half. . . . High productivity of labor depended above all on continuity in the production of the same type [of vessel]. (*Ships For Victory*, pp. 202, 232, and 826.)

Shipyards reorganized work to enable workers to benefit from the learning effect. Jobs were broken down into simpler tasks and each worker was assigned to one of these tasks:

> When the work was thus planned so that the same crew had the same task every day, there was no need of teaching the new workmen the 40 to 70 operations which had formed part of the craft learned by an apprentice . . . Instead a man who had only one skill could be kept busy doing that one thing. . . . a workman specialized on just one kind of work . . . and since he concentrated on one operation he could learn that one quickly. (*Ships For Victory*, pp. 237–8.)

This basic principle of the division of labour had been stated by Adam Smith in 1776: 'the division of labour, by reducing every man's business to some one simple operation, and by making this operation the sole employment of his life, necessarily increases very much the dexterity of the workman.' (*An Inquiry into the Nature and Causes of the Wealth of Nations*, 1776, Book I, Chapter I.)

This principle was further developed by the engineer-mathematician Charles Babbage in 1835. Babbage argued that, through the division of labour, unit labour cost could be reduced by two distinct methods.

First, 'by dividing the work to be executed into different processes, each requiring different degrees of skill or of force, [the employer] can purchase exactly that precise quantity of both which is necessary for each process; whereas, if the whole work were executed by one workman, that person must possess sufficient skill to perform the most difficult, and sufficient strength to execute the most laborious, of the operations into which the art is divided.'

Thus the division of labour permits skilled-labour-time to be obtained and used in exactly the quantity required and applied to only the tasks requiring skill. Conversely, cheaper unskilled-labour-time could be applied to tasks not requiring skill, thus ensuring that expensive skilled-labour-time was not 'wasted' on tasks not requiring skill.

Second, 'The constant repetition of the same process necessarily produces in the workman a degree of excellence and rapidity in his particular department, which is never possessed by a person who is obliged to execute many different processes.' (Charles Babbage, *On the Economy of Machinery and Manufactures* (1835; Reprinted Edition, William Pickering, 1989), pp. 125 and 123; see also the discussion of the learning effect by Alfred Marshall, *Principles of Economics* (First Edition, 1890), Book IV, Chapter IX.) This second method noted by Adam Smith and Charles Babbage is simply the learning effect.

The principles we are discussing in relation to the learning effect are thus well-known principles of long standing, both in economic theory and in scientific management. The simplification of jobs speeded up the learning effect by enabling each worker to concentrate on one repetitive task and to become highly proficient in that one task: this was a lesson of 'scientific management' which had long previously been known to the automobile industry in its utilization of labour. These principles were utilized by shipyards building Liberty ships.

Table 6.3 shows the data collected and published on unit labour requirement in just one shipyard, identified as Yard F, which built a total of 173 Liberty ships.

The data are published according to the 'batch' of ships in order of delivery

Table 6.3
The learning effect in Yard F building Liberty ships, United States

Vessel nos. (in order of delivery dates), batch	Thousand labour-hours: Total labour-hours used,[a] or batch total labour-hours	Average unit labour requirement,[a] 'batch average'	Index of batch average unit labour requirement, batch 1 to 10 equals 100
1 to 10	12,170	1,217	100
11 to 20	9,730	973	80
21 to 30	9,240	924	76
31 to 40	9,490	949	78
41 to 50	7,500	750	62
51 to 60	7,180	718	59
61 to 70	7,000	700	58
71 to 80	6,610	661	54
81 to 90	7,100	710	58
91 to 100	6,900	690	57
101 to 110	6,520	652	54
111 to 120	6,200	620	51
121 to 130	5,930	593	49
131 to 140	5,770	577	47
141 to 150	5,500	550	45
151 to 160	5,450	545	45
161 to 170	5,300	530	44
171 to 173[b]	1,575	525	43

(*a*) For that batch, or group, of vessels.
(*b*) Yard F built 173 Liberty ships. The data relate to ships delivered through to 31 October 1944.

Source: United States Department of Labor Bureau of Labor Statistics, *Wartime Employment, Production, and Conditions of Work in Shipyards*, Bulletin No. 824, 1945, table 23, pp. 40–41.

dates. Thus vessels no. 1 to no. 10 inclusive can be regarded as the first batch (of ten) vessels to be built by Yard F, vessels no. 11 to no. 20 inclusive can be regarded as the second batch (of ten) vessels to be built by Yard F, and so on. The total labour requirement, measured in labour-hours needed to build each batch (of ten) vessels, is given in the column 'Total labour-hours used' (note, however, that the last batch comprised only three vessels).

The unit labour requirement – the labour-time required to build one Liberty ship – may then be calculated, not for each individual vessel but averaged over the first batch, over the second batch, and so on. Each of these averages of unit labour requirement we may call the 'batch average'.

The first ten Liberty ships built by Yard F required on average 1,217,000 labour-hours per vessel. But the second group of ten Liberty ships built by Yard F required on average only 973,000 labour-hours per vessel. The batch average for the second batch was 20 per cent lower than the batch average for the first batch.

As we follow the batch averages, we see a continuing decrease, interrupted on only two occasions (and then only slightly). The last three Liberty ships to be built in Yard F had a unit labour requirement, on average, of 525,000 labour-hours per vessel. This is 57 per cent (!) lower than the batch average for the first batch.

The important point about Table 6.3 is that it demonstrates the substantial and sustained impact which the learning effect can have on unit labour requirement (or, conversely, on labour productivity) in a sizeable plant.

Because the data of Table 6.3 are not reported by period of elapsed time, we are not able to present a progress function for Yard F. Furthermore, we cannot present straightforwardly a learning curve for Yard F because the batch average unit labour requirements are reported with respect to a batch of arbitrary size and it is a problematic matter to relate such batch averages to the cumulative total of output produced. The problem here is analogous to that discussed for the assembling of propelling pencil mechanisms in Table 6.1.

Nevertheless, a learning curve can be derived both for the production of Liberty ships and for the assembling of propelling pencil mechanisms, and this derivation will be explained in the following chapter.

It is important to note that the learning effect was not confined to Yard F: all shipyards building Liberty ships benefited from the learning effect.

Table 6.4 shows the percentage reduction, between the first batch average (in each case for the first batch of 10 vessels) and the last batch average (in each case for whatever number of vessels comprised the last batch), in unit labour requirement in sixteen shipyards building Liberty ships (including Yard F).

At this point in the exposition, we have a difficulty in comparing, in any straightforward way, the improvement in performance among the various shipyards because each built a different number of vessels in total and so each shipyard had a different amount of experience over the whole length of

Table 6.4
The substantial and sustained impact of the learning effect in sixteen shipyards

Yard	Number of vessels built[(a)]	Batch average unit labour requirement, thousand labour-hours		Percentage reduction in batch average unit labour requirement, first to last
		First batch	Last batch	
A	384	1,024	450	–56
B	351	1,009	364	–64
C	330	899	464	–48
D	306	1,241	456	–63
E	186	1,395	427	–69
F	173	1,217	525	–57
G	138	915	562	–39
H	126	740	401	–46
I	110	1,164	633	–46
J	72	1,342	697	–48
K	65	1,837	601	–67
L	61	1,667	553	–67
M	56	1,571	570	–64
N	20	1,057	889	–16[(b)]
O	15	1,528	1,095	–28[(b)]
P	11	2,347	1,404	–40[(b)]

(*a*) Through to 31 October 1944; excludes all modified Liberty ships.
(*b*) Only two batches.

Source: as for Table 6.3.

its production run. Subsequently, when using the learning curve, we can show how such comparisons may best be made – the 'hidden agenda' of Table 6.4 is to enable you subsequently to appreciate the usefulness of the learning curve in such comparisons.

Excluding Yards N, O, and P, which built 20 or fewer vessels and so had much less experience than did the other shipyards, the percentage reduction in each case is substantial, ranging from a low of –39 per cent to a high of –69 per cent.

The fact that in all shipyards unit labour requirement steadily decreased meant steadily falling unit cost per vessel, and this caused considerable problems for the Maritime Commission in setting prices in its 'cost-plus contracts' with the shipyards. A constant price for the construction of vessels would have meant considerably increased profits for the (privately owned) shipyards, and 'profiteering' was a sensitive issue in wartime. The Maritime Commission solved the problem piecemeal by negotiating, through its labour-hour cost contracts, reduced prices per vessel based on what it thought should be the reduction in unit labour requirement.

Each successive contract (with the same yard for successive batches of Liberty ships) had its cost estimate, based on (forecast) unit labour requirement, revised downwards as the Maritime Commission tried to prevent 'excessive' profits being made per vessel.

For example, in three shipyards for which there are published data for successive contracts, the following were the respective contract estimates for labour-hours per Liberty ship (*Ships For Victory*, table 10, p. 475):

	Contracted labour-hours per vessel		
Yard	*Contracts awarded, December 1942*	*Contracts awarded, April–June 1943*	*Percentage change*
Houston Shipbuilding Corpn, Houston	640,700	564,700	–12
Oregon Shipbuilding Corpn, Portland	634,000	558,900	–12
Bethlehem-Fairfield Yard, Baltimore	622,600	548,800	–12

The basis for the contracted fee (effectively, for the 'price paid' per vessel) was thus successively negotiated downwards to ensure that the Maritime Commission (acting for the nation and under the vigilant scrutiny of Congress) received, per vessel, a large part of the benefit of the learning effect. Even with such negotiation, there was, more than occasionally, a retrospective 're-pricing' of contracts when, in a shipyard, the Maritime Commission's resident auditors/cost-accountants found actual costs to be lower than expected (and profits therefore higher than initially forecast).

All this is evidence of how the learning effect was persistently and systematically acting so as to decrease unit cost and thence to increase profit.

However, all this negotiating of contracts and retrospective 're-pricing' seems to have been done on an *ad hoc* basis through a process of bargaining with individual yards on a piecemeal basis. If the Maritime Commission had known about the learning curve, then the pricing of contracts for successive batches could have been done much more systematically using the learning curve. The next chapter explains the learning curve.

7 The Learning Curve

7.1 The cumulative total labour requirement function

A learning curve may be defined as a function which relates performance to experience. In the case of production economics, the learning curve may use one of several possible measures of unit labour requirement as a measure of performance, and cumulative total output is used as a measure of experience. The learning curve may be extended to use unit cost as a measure of performance and this extension will be discussed at the end of the chapter.

In the example of assembling propelling pencil mechanisms (Table 6.1), unit labour requirement was reported as period average unit labour requirement. In the example of Yard F building Liberty ships (Table 6.3), unit labour requirement was reported as batch average unit labour requirement. Neither of these can straightforwardly be related to cumulative total output because each is given with respect to an arbitrary coverage (week for propelling pencil mechanisms, batch of ten vessels for Liberty ships).

The only way to avoid arbitrary coverage would be to have the unit labour requirement reported on an item-by-item basis, and data are hardly ever recorded and reported in this way.

Nevertheless, it is possible to derive an exact form of the learning curve which gives an item-by-item matching between performance and cumulative total output. In order to do this we must start from what I call the 'cumulative total labour requirement function'.

The cumulative total labour requirement function is the fundamental basis for any learning curve analysis.

To explain the cumulative total labour requirement function consider first the mechanics of the decrease in batch average unit labour requirement. Consider the behaviour of any series of batch averages where production is subject to the learning effect so that the batch average unit labour requirement decreases successively. (The following argument could equally well be developed in respect of period averages, so it is a completely general argument.)

In terms of arithmetic, what are the 'mechanics' of the decrease in batch average unit labour requirement as we go from one batch to the next? The answer is simple. In order for there to be a decrease in the batch average unit labour requirement the cumulative total of labour-hours used must increase at a *lower* proportionate rate than does the cumulative total of output produced.

This point is very important and basic and is the fundamental starting point for deriving the learning curve. To illustrate in the simplest way possible, suppose that 1,000 labour-hours are used to produce the first batch of 100 items. Then the batch average unit labour requirement is 10 labour-hours per item. If 800 labour-hours are used to produce the second batch of 100 items, then the batch average unit labour requirement decreases to 8 labour-hours per item.

In terms of arithmetic, what has happened in going from the first batch to the second batch is that the cumulative total of labour-hours used has increased by 80 per cent (from 1,000 cumulative total labour-hours by the end of the first batch to 1,800 cumulative total labour-hours by the end of the second batch). But cumulative total output has increased by 100 per cent (from 100 items of cumulative total output by the end of the first batch to 200 items of cumulative total output by the end of the second batch).

This sort of relationship between the cumulative total of labour-hours used and the cumulative total of output produced is essential if the batch average unit labour requirement is to decrease. You can appreciate that if cumulative total labour-hours also increased by 100 per cent, to 2,000 labour-hours by the end of the second batch, then producing the second batch would have required 1,000 labour-hours, and the batch average unit labour requirement for the second batch would have remained constant at 10 labour-hours per item.

We may apply this analysis to the data for Yard F producing Liberty ships (Table 6.3). First, we have the total number of labour-hours used in producing each batch of vessels. Each of these may be called the 'batch total labour-hours'. These figures are given in the first column of data in Table 7.1; note that the last batch relates to a batch of different size than the others.

Second, we calculate the cumulative total of labour-hours used simply as a running total of the batch total labour-hours. These figures are given in the second column of Table 7.1. To illustrate, 21,900 is calculated as 12,170 + 9,730; 31,140 is calculated as 12,170 + 9,730 + 9,240. And so on, proceeding always in sequence. Observe that the figure 21,900 thousand labour-hours is the cumulative total labour-hours used in the production of vessels no. 1 to no. 20 inclusive.

Third, cumulative total output may be read in Table 7.1 as the second number in each batch of vessel numbers. Thus, by the end of the first batch,

Table 7.1
The relationship between cumulative total labour-hours used and cumulative total output produced, Yard F

	Thousand labour-hours		Index, first batch = 100	
Batch, vessel nos.[(a)]	Total labour-hours used, or batch total labour-hours[(b)]	Cumulative total labour requirement, measured in labour-hours used,[(c)] $CTL_{(i)}$	Cumulative total labour-hours used	Cumulative total output produced
1 to 10	12,170	12,170	100	100
11 to 20	9,730	21,900	180	200
21 to 30	9,240	31,140	256	300
31 to 40	9,490	40,630	334	400
41 to 50	7,500	48,130	395	500
51 to 60	7,180	55,310	454	600
61 to 70	7,000	62,310	512	700
71 to 80	6,610	68,920	566	800
81 to 90	7,100	76,020	625	900
91 to 100	6,900	82,920	681	1,000
101 to 110	6,520	89,440	735	1,100
111 to 120	6,200	95,640	786	1,200
121 to 130	5,930	101,570	835	1,300
131 to 140	5,770	107,340	882	1,400
141 to 150	5,500	112,840	927	1,500
151 to 160	5,450	118,290	972	1,600
161 to 170	5,300	123,590	1,016	1,700
171 to 173	1,575	125,165	1,028	1,730

(*a*) The second number in this column gives a sample of values of *i* (namely, those values of *i* reached by the end of that particular batch); accordingly these values of *i* can be used in the bracketed subscript for the cumulative total labour requirement, $CTL_{(i)}$.

(*b*) This is the batch average unit labour requirement from Table 6.3 multiplied by 10, the number of vessels in the batch, except in the case of the last batch when the multiplication is by 3, Yard F having built 173 vessels in total, so the last batch comprised only 3 vessels.

(*c*) For all vessels up to and including the last number in the sequence; calculated as the running total of the preceding column.

Source: Table 6.3.

10 vessels measures the cumulative total output; by the end of the second batch, 20 vessels measures the cumulative total output; and so on.

The learning effect expresses itself in the arithmetical relationship that the cumulative total of labour-hours used increases at a lower proportionate rate than does the cumulative total output. This may be more readily seen if we express each of cumulative total labour-hours used and cumulative total output produced in index form with a base of 100, as in the last two columns of Table 7.1. We can see that, throughout, the proportionate increase in cumulative total labour-hours used is less than the proportionate increase in cumulative total output produced.

In order to analyse this basic relationship in a given plant between cumulative total labour requirement and cumulative total output we use the following notation:

i = 1, 2, 3, ..., n, denoting the serial number of the item of product in sequence of making in the given plant; n is the total number of items which will have been made when the production run comes to an end; it is important to note that i must be a positive integer; the sequence i refers to items (of a 'model' of a product) *ever* made by the given plant; it is also presumed that the sequence i proceeds in an uninterrupted sequence.

UL_i = The quantity of labour-time required to make the ith item in the given plant; note carefully that the subscripted term UL_i denotes the unit labour requirement for a particular single item of product in the sequence of production; in recognition of this we shall call it a 'particular item unit labour requirement' (this distinguishes it from the period average or batch average unit labour requirement); for instance, UL_1 should be read in full as: *U*nit *L*abour requirement for the production of the first item of output to be produced (of that particular product from the given plant); UL_{11} should be read as: *U*nit *L*abour requirement for the production of the eleventh item of output to be produced (of that particular product from the given plant).

$CTL_{(i)}$ = Cumulative total labour requirement to produce i items in the given plant; that is:

$$CTL_{(i)} \equiv \sum_{j=1}^{j=i} UL_j$$

Note that we use the bracketed subscript (i) to denote the *cumulative* total from the start of production up to and including i; thus $CTL_{(i)}$ should be read as: *C*umulative *T*otal *L*abour requirement totalled over the production of items 1 to i inclusive.

c = A parameter to be called the 'elasticity of cumulative total labour requirement with respect to cumulative total output', or the 'cumulative total labour requirement elasticity' for short; c is the basic parameter relating to the learning effect.

We have seen that the learning effect results in a relationship wherein the proportionate increase in cumulative total labour requirement (measured, say, in cumulative total labour-hours used) is less than the proportionate increase in cumulative total output produced.

In order to explore this relationship we may write cumulative total labour requirement, $CTL_{(i)}$, as a power function of cumulative total output produced, i, as follows:

$$CTL_{(i)} = UL_1 \; i^c$$

This function I call the 'cumulative total labour requirement function'.

In the first place, the function says that the cumulative total labour requirement (cumulative total labour-hours) required for the first item is equal to the particular item unit labour requirement for item 1, because, of course, when $i = 1$ we have $i^c = 1^c = 1$.

In the second place, the function says that, thereafter, $CTL_{(i)}$ varies with i, the cumulative total output produced, but may not vary in equal proportion.

That $CTL_{(i)}$ increases with i but perhaps at a lower proportionate rate can be expressed in the value of the parameter c, the elasticity of the cumulative total labour requirement with respect to cumulative total output. The value of c must be positive, indicating that as cumulative total output increases so too must cumulative total labour requirement increase.

Furthermore, and this is the important point, if learning is occurring the value of c must be less than 1. If $c = 1$, then $CTL_{(i)}$ varies equi-proportionately with i. In such a case it is clear that no learning is occurring and batch average (or period average) unit labour requirement will remain constant. That is, if $c = 1$, then $CTL_{(i)}$ will simply be a multiple, i (= $i^c = i^1 = i$), of UL_1.

But if $c < 1$ (and is positive), then the proportionate increase in $CTL_{(i)}$ will be *lower* than the proportionate increase in i.

For example, if $c = 0.8$, then $CTL_{(i)}$ will be a multiple $i^c = i^{0.8}$ of UL_1 and $i^{0.8}$ will always be less than i. For example, $CTL_{(2)}$ will be UL_1 multiplied by $2^{0.8} = 1.741$, so that cumulative total labour requirement will increase by only 74.1 per cent when cumulative total output increases by 100 per cent (or doubles) from $i = 1$ to $i = 2$.

Accordingly, the learning effect means that $0 < c < 1$, and the smaller is c, the greater the learning effect, because the smaller is the cumulative total labour requirement elasticity, the less will be the proportionate increase in cumulative total labour requirement for any given proportionate increase in cumulative total output.

The magnitude of the cumulative total labour requirement elasticity thus summarizes, and measures, the magnitude of the learning effect over a whole set of data.

The cumulative total labour requirement function is the basic underlying theoretical function on which all learning curve analysis is based and from which all the learning concepts and functions are derived.

It follows from this that the cumulative total labour requirement elasticity, the parameter c, is the basic underlying theoretical parameter in learning curve analysis. For this reason, I explain and identify all other parameters relating to the learning effect in terms of c.

The practical importance of the cumulative total labour requirement function should also be emphasized because, as we shall show in this chapter, it can be applied both to data which are recorded and reported according to batches and also to data which are recorded and reported according to periods of elapsed time.

To illustrate the cumulative total labour requirement function for a given plant, consider the data in Table 7.1 for Yard F. In Table 7.1, i is successively the second number in each batch (group of vessels); that is, we have $i = 10, 20, ..., 170, 173$. In the case of Yard F, $n = 173$. The fact that we do not have observations on all values of i ($i = 1, 2, 3, 4, ..., 173$) is not of great importance; it simply means that we have a sample of 18 observations instead of the full set of 173 observations, but we can work almost as well with a sample (providing the sample is sufficiently large for statistical purposes).

In Table 7.1, $CTL_{(i)}$ is the cumulative total labour requirement (measured in labour-hours used) and is given in the second column of data. For instance, measured in labour-hours, $CTL_{(10)}$ is 12,170,000; $CTL_{(20)}$ is 21,900,000; and so on.

The cumulative total labour requirement function for Yard F is depicted in Figure 7.1, using double-logarithmic scales because the power function is linear in logarithms.

Figure 7.1 plots cumulative total labour requirement on the logarithmic scale vertical axis against cumulative total output on the logarithmic scale horizontal axis.

By using this sample of observations and the method of least-squares applied to the logarithms of the variables, we may estimate the cumulative total labour requirement function for Yard F to have the following empirical value:

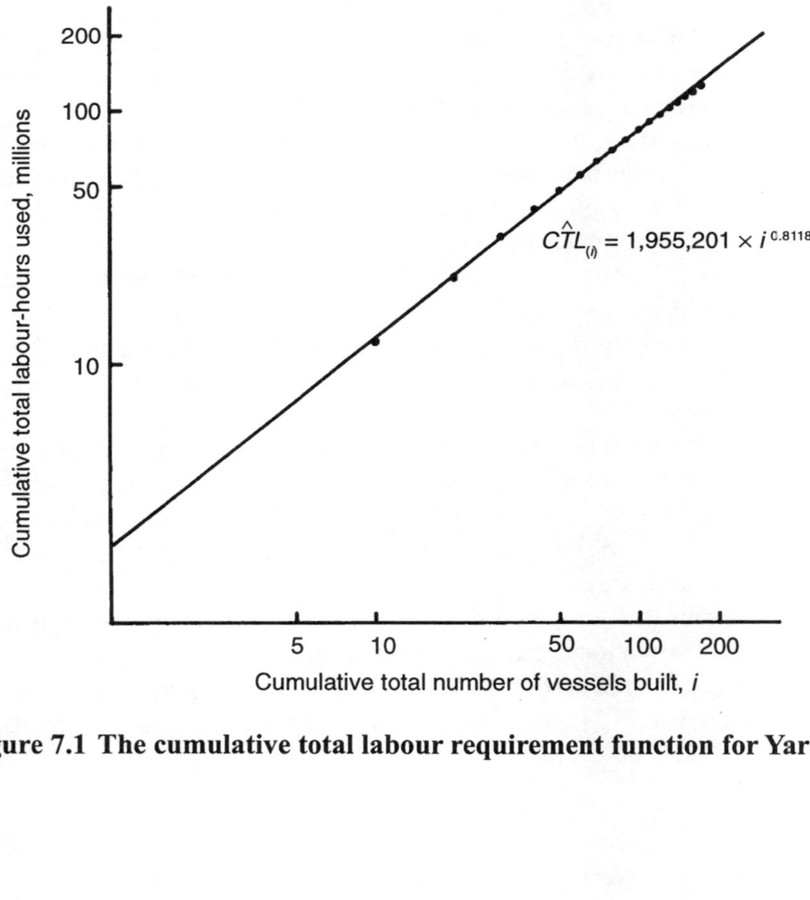

Figure 7.1 The cumulative total labour requirement function for Yard F

$$\hat{CTL}_{(i)} = 1{,}955{,}201 \times i^{0.8118}$$

with a correlation coefficient of 0.9995, indicating that this power function, linear in logarithms, provides a close fit to the observed points. (Here, as previously, a hat (^) placed above a variable denotes an estimated value of that variable as opposed to an actual observed value of that variable.)

The value 1,995,201 is an estimate of the labour-hours required by Yard F to build its first Liberty ship.

For Yard F the elasticity of the cumulative total labour requirement with respect to cumulative total output is 0.8118, which is positive and considerably less than 1. The value of this parameter summarizes, over the whole set of data for Yard F, the impact of the learning effect, and, for example, enables us to say that, for each doubling of cumulative total output, cumulative total labour requirement increases by $2^{0.8118} = 1.755$, or 75.5 per cent. (This is, of course, a statement based on a statistical estimate, and so the statement relates to a general statistical *tendency* rather than to each and every *actual* proportionate change in cumulative total labour requirement which may be observed in Table 7.1 when cumulative total output doubles.)

The cumulative total labour requirement function is not a learning curve as such, because one would not normally take cumulative total labour input as a measure of performance. However, as we shall show, the cumulative total labour requirement function is the basis for deriving the various learning curves which are used in analysing the learning effect.

7.2 The cumulative average unit labour requirement learning curve

One important measure of performance associated with the learning curve is cumulative average unit labour requirement. The cumulative average unit labour requirement is calculated by dividing the cumulative total labour requirement by the cumulative total output produced.

Table 7.2 illustrates for Yard F how the cumulative average unit labour requirement is calculated. The first column (effectively, the row headings) gives i, the cumulative total output measured in number of vessels ever built (of that particular type of vessel by that particular shipyard). The second column gives $CTL_{(i)}$, the cumulative total labour requirement, here measured in labour-hours used.

When we divide the cumulative total labour requirement by the cumulative total output produced, we get the cumulative average unit labour requirement,

Table 7.2
The cumulative average unit labour requirement, Yard F

	Thousand labour-hours	
Cumulative total output, i (in vessels)	Cumulative total labour requirement, measured in labour-hours, $CTL_{(i)}$	Cumulative average unit labour requirement, $CAL_{(i)} \equiv CTL_{(i)}/i$
10	12,170	1,217.000
20	21,900	1,095.000
30	31,140	1,038.000
40	40,630	1,015.750
50	48,130	962.600
60	55,310	921.833
70	62,310	890.143
80	68,920	861.500
90	76,020	844.667
100	82,920	829.200
110	89,440	813.091
120	95,640	797.000
130	101,570	781.308
140	107,340	766.714
150	112,840	752.267
160	118,290	739.313
170	123,590	727.000
173	125,165	723.497

Source: Table 7.1.

shown in the third column. This must be understood carefully as a running average over all the vessels so far produced. The cumulative average will be greater than the batch average because cumulative average is influenced by the 'tail' of data back to and including the first item ever produced (of that particular model of product from the given plant).

The behaviour of the cumulative average unit labour requirement is illustrated in relation to cumulative total output in Figure 7.2. As was not the case in Figure 6.3, we have in Figure 7.2 an exact or non-arbitrary match between each cumulative average unit labour requirement and its corresponding cumulative total output. This happens, of course, because each cumulative average is calculated with respect to its corresponding cumulative total.

To illustrate this feature explicitly, Figure 7.2 is drawn with horizontal lines, the *height* of each of which represents each cumulative average, with the coordinate matching that cumulative average to the cumulative total output produced given as a point at the rightwards end of the horizontal line.

The learning curve will then be the curve drawn through the points at the end of each horizontal line.

The important point to note is that if the cumulative averages were to be recorded at points other than the ones shown in Figure 7.2 they would fall on the same curve (although at different points along that curve). In other words, it makes no difference to the position of this curve just where each cumulative average, and its corresponding cumulative total output, happens to be recorded.

To illustrate this, I have indicated, each with a broken line and a cross, where the cumulative average unit labour requirement would be for a cumulative total output of 15 vessels and for a cumulative total output of 25

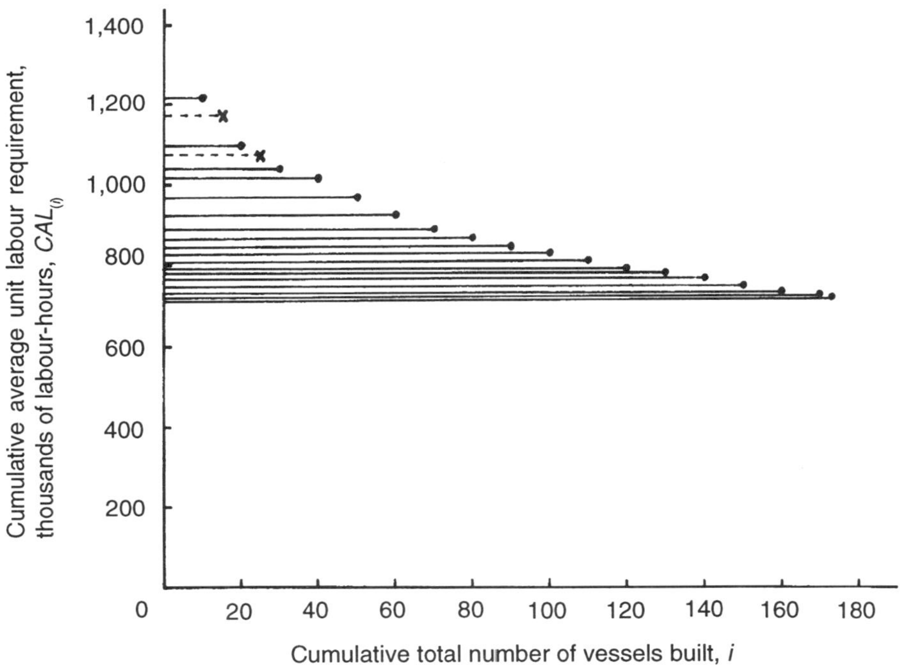

Figure 7.2 Cumulative average unit labour requirement learning curve for Yard F

vessels (these points having been estimated by using the empirical cumulative total labour requirement function). We can see that these (estimated) points would lie pretty much on the same curve as that traced out by the actual observations.

There is thus nothing arbitrary about the coordinates in Figure 7.2; the position of the curve does not depend on where the observations are recorded and reported. The same would be true for any set of observations on cumulative average unit labour requirement by period instead of by batches.

Consequently, Figure 7.2 can be said to depict a learning curve and so may properly be called such. Because it uses the cumulative average unit labour requirement as a measure of performance, I call it the 'cumulative average unit labour requirement learning curve'.

We use the following additional notation:

$CAL_{(i)}$ = Cumulative average unit labour requirement up to and including the *i*th item produced in the given plant; $CAL_{(i)}$ should not appear without the cumulative (*i*) subscript, and should be read in full as: *C*umulative *A*verage unit *L*abour requirement averaged over the production of items 1 to *i* inclusive.

The definitional equation for calculating the cumulative average unit labour requirement is (see Table 7.2):

$$CAL_{(i)} \equiv CTL_{(i)}/i$$

To derive the cumulative average unit labour requirement learning curve we may substitute the cumulative total labour requirement function, as follows:

$$CAL_{(i)} = UL_1 \, i^c / i$$

$$= UL_1 \, i^{c-1}$$

This is the expression for the cumulative average unit labour requirement learning curve. (It was more or less in this form that the learning curve was first introduced in the pioneering article by T.P. Wright, 'Factors Affecting the Cost of Airplanes', *Journal of the Aeronautical Sciences*, vol. 3, no. 4, February 1936. Wright's words (p. 124 and see fig. 1) are, with my interpolated interpretations: 'A curve may be plotted which shows directly the relationship between ... the average labor cost in any quantity [that is, the unit labour cost equivalent of cumulative average unit labour requirement] . . . [and] production quantity [that is, cumulative total output]'. Wright also presented (fig. 2) another form of the learning curve which we shall discuss shortly.)

We can see that the parameters in the cumulative average unit labour requirement learning curve, namely UL_1 and $c - 1$, are from or are based on the cumulative total labour requirement function. Thus by starting from the cumulative total labour requirement function we can derive one exact form of the learning curve. (Note that in much of the literature the parameter $c - 1$ is often designated by a single letter and may be referred to as the 'slope of the learning curve'. However, given my opinion on the primacy of the cumulative total labour requirement function, and hence on the primacy of c, I prefer throughout to work in terms of c, the cumulative total labour requirement elasticity. Furthermore, as we shall see, c is the parameter which we need when we apply the binomial theorem to the derivation of the unit labour requirement learning curve.)

If learning is occurring, then $0 < c < 1$; consequently the exponent in the cumulative average unit labour requirement learning curve, $c - 1$, must be negative.

Figure 7.2 is drawn on arithmetic scale graph paper (which is a common form of presentation of such learning curves), but we can see that the cumulative average unit labour requirement learning curve is linear in logarithms. That is, if we take the logarithm of both sides we obtain the equation for a straight line, as follows:

$$\text{logarithm } CAL_{(i)} = \text{logarithm } UL_1 + (c - 1) \text{ logarithm } i$$

This equation can be used together with the statistical technique of least-squares to estimate the empirical values for any particular cumulative average unit labour requirement learning curve.

For Yard F and the data in Table 7.2, we have (taking anti-logs of the estimated values so as to get back to the original power function):

$$\hat{CAL}_{(i)} = 1{,}955{,}200 \times i^{-0.1882}$$

with a correlation coefficient of 0.9915.

The constant term is virtually identical to the constant term in the cumulative total labour requirement function (the difference arises merely from using, in Table 7.2, an insufficient number of decimal places in the cumulative average unit labour requirement).

We observe that the exponent is equal to the previously estimated cumulative total labour requirement elasticity minus 1; that is, 0.8118 – 1 = –0.1882.

The exponent in the cumulative average unit labour requirement learning curve, $c - 1$, may be referred to in full as the elasticity of the cumulative average unit labour requirement with respect to cumulative total output, and this lengthy term may be abbreviated to 'learning curve elasticity'; but we shall always write it as $c - 1$ and shall not designate it by a single letter.

A common way of using the learning curve elasticity is to refer to the decrease in cumulative average unit labour requirement when cumulative total output doubles. There is no reason for taking a doubling of cumulative total output other than that a doubling is an intuitively easy factor of increase to understand. The following could just as well apply to a factor of increase of 3, but an increase by a factor of 2 is the established convention. Taking a doubling, we have:

$$CAL_{(i)} = UL_1\, i^{c-1} \quad \text{and} \quad CAL_{(2i)} = UL_1\, (2i)^{c-1}$$

Whence:

$$CAL_{(2i)} \,/\, CAL_{(i)} = 2^{c-1}$$

Whence:

$$CAL_{(2i)} = 2^{c-1} \times CAL_{(i)}$$

The convention in the learning curve literature is to take $(2^{c-1})100$ and to refer to this as a '$(2^{c-1})100$ per cent learning curve'.

For example, the learning curve for Yard F would be referred to as an '87.77 per cent learning curve' (because $2^{0.8118-1} \times 100 = 87.77$ per cent).

I shall refer to any empirical value calculated as $(2^{c-1})100$ as the 'learning curve rate' and shall always give the learning curve rate as a percentage to avoid confusing it with other parameters relating to the learning effect. The learning curve rate gives the change in unit labour requirement resulting from a doubling of cumulative total output, the change being expressed by giving the decreased unit labour requirement as a percentage of the initial unit labour requirement.

The usefulness of the cumulative average unit labour requirement learning curve is that if one has an estimate of the unit labour requirement for the first item to be produced and an estimate of the learning curve elasticity, then one may estimate the cumulative average unit labour requirement for a production run of any particular length and this may be used as a basis for estimating the unit labour cost for a production run of that length.

In his 1936 article, Wright concluded with a demonstration of how the learning curve could be applied to the problem of estimating the likely production cost for a 'programme sponsored by the Bureau of Air Commerce for the development of a small two-place airplane . . . assuming a quantity of ten thousand units could be released for construction'.

7.3 The particular item unit labour requirement learning curve

Another form of the learning curve was also introduced by Wright in his 1936 article (see the top part of fig. 2 in that article). This uses, as a measure of performance, the unit labour requirement for the *i*th item, UL_i, or in Wright's words and with Wright's substitution of unit labour cost for unit labour requirement, a curve to plot 'the cost of each machine [item] of a series'.

However, Wright considered that this would be difficult to do 'because of the greater difficulty in securing reliable empirical information on the relative cost of each machine [item] of a series, since accounting methods seldom reveal such data'. Unfortunately, Wright, and many others since, failed to realize that the learning curve parameters not only could but also should be estimated from the cumulative total labour requirement function, which does not require information to be recorded and reported on an item-by-item basis.

I now explain the derivation of the learning curve which uses the unit labour requirement for a particular *i*th item, UL_i, as a measure of performance. The derivation of this form of the learning curve is more complex than is usually recognized. This form of the learning curve may be used in an exact formula (which is complex but which may be simplified) or, much more commonly, in an approximate formula (which is less complex and of which there are two variants).

We begin with the cumulative total labour requirement function. The unit labour requirement for any *i*th item may be expressed as the difference between the cumulative total labour requirement for *i* and the cumulative total labour requirement for $i - 1$:

$$UL_i = CTL_{(i)} - CTL_{(i-1)}$$

Substitute the cumulative total labour requirement function to get:

$$\begin{aligned} UL_i &= UL_1\, i^c - UL_1\, (i-1)^c \\ &= UL_1(i^c - (i-1)^c) \end{aligned}$$

This is the exact formula for UL_i.

Although the exact formula may be used quite readily, it may be simplified by further factorization and rearrangement as follows:

$$\begin{aligned} UL_i &= UL_1\, i^c\,(1-(i-1)^c/i^c) \\ &= UL_1\, i^c\,(1-[(i-1)/i]^c) \\ &= UL_1\, i^c\,(1-[1-(1/i)]^c) \end{aligned}$$

I shall refer to this simplified form of the exact formula as the 'simplified exact formula' for UL_i. Whenever it is wished to obtain the most accurate estimate of UL_i the simplified exact formula should be used (and, of course, the exact formula can also be used because each will give exactly the same result).

Because the exact formula and the simplified exact formula each has UL_i as the dependent variable I call each of these, and any other formula which expresses UL_i as a function of i, the 'particular item unit labour requirement learning curve'. This is cumbersome terminology but it draws a clear distinction with the cumulative average unit labour requirement learning curve.

The simplified exact formula can be approximated by a less complex formula which is useful for statistical analysis. This less complex formula is commonly identified as *the* learning curve (following Wright). This approximate formula has two variants. The derivation of the first variant may be explained as follows.

The approximation involves considering the expression $[1-(1/i)]^c$ and simplifying this. This simplification involves using the binomial theorem (an important mathematical theorem due to Isaac Newton). According to one version of Newton's binomial theorem, an expression such as $(1+x)^c$, where c is not a positive integer, may be written or expanded as follows:

$$\begin{aligned} (1+x)^c &= 1+cx+\frac{c(c-1)}{2\times 1}x^2 \\ &\quad +\frac{c(c-1)(c-2)}{3\times 2\times 1}x^3 \\ &\quad +\ldots+\frac{c(c-1)\,(c-2)\ldots(c-r+1)}{r!}x^r+\ldots \end{aligned}$$

where $r!$, read as 'r factorial', is defined as the product of all the positive integers between r and 1 inclusive, r always being a positive integer.

In the case we are considering, $x = -1/i$ and c is a positive fraction, so the series, although infinite, 'converges' to a limiting value. Making the substitution for x we may write:

$$[1 - (1/i)]^c = 1 + c(-1/i) + \frac{c(c-1)}{2!} \times (-1/i)^2$$

$$+ \frac{c(c-1)\,(c-2)}{3!} \times (-1/i)^3 + \ldots$$

Simplifying and keeping a careful watch on the signs involved, we get:

$$[1 - (1/i)]^c = 1 + (-c/i) + \frac{c(c-1)}{2!} \times \left(\frac{1}{i^2}\right)$$

$$+ \frac{c(c-1)\,(c-2)}{3!} \times \left(-\frac{1}{i^3}\right) + \ldots$$

For large values of i we can ignore the third and all subsequent terms on the right-hand side of the preceding equation because these involve powers of i in the denominator and are therefore relatively small.

Thus a reasonable approximation for $[1-(1/i)]^c$, especially when i is large, may be obtained by using only the first two terms on the right-hand side of the preceding equation, namely:

$1 + (-c/i)$ or $1 - (c/i)$

Using this result in the simplified exact formula we may write (and strictly speaking, we should use the approximation sign, ≈, instead of the equals sign, but the approximation is likely to be sufficiently close to permit the use of the equals sign providing this qualification is taken as understood):

$$\begin{aligned} UL_i &= UL_1\, i^c\, (1 - [1 - (1/i)]^c) \\ &= UL_1\, i^c\, (1 - (1 - (c/i))) \\ &= UL_1\, i^c\, (c/i) \\ &= UL_1\, c\, i^{c-1} \end{aligned}$$

This is the first variant of the approximate formula for the particular item unit labour requirement learning curve. This is the most commonly used form of the learning curve, and is commonly identified simply as 'the learning curve' because it uses the readily understandable unit labour requirement for the *i*th item, UL_i, as a measure of performance and has *i* alone as the independent variable on the right-hand side.

Note that the particular item unit labour requirement learning curve has the same parameters as occur in the cumulative total labour requirement function. Note especially that the constant term is UL_1c.

The approximate formula for the particular item unit labour requirement learning curve has the same elasticity as for the cumulative average unit labour requirement learning curve, namely, $c - 1$.

Consequently, the same result applies whenever cumulative total output is doubled; that is: $UL_{2i}/UL_i = 2^{c-1}$. However, this may apply only as an approximation, especially if small values of *i* are involved.

There are two variants of the approximate formula. The formula just derived is the first variant and it holds as a good approximation for large values of *i*.

For small values of *i*, a better approximation may be derived and this is the second variant. This second variant is due to A.B. Berghell, *Production Engineering in the Aircraft Industry* (McGraw-Hill, 1944), Chapter XII Learning Curves. Berghell's derivation of the second variant is as follows.

Let the approximate formula for the learning curve for UL_i when values of *i* are small, be:

$$UL_i \quad = \quad UL_1\, c(i + x)^{c-1}$$

where *x* is a constant whose value is to be determined.

The exact formula for UL_i is:

$$UL_i \quad = \quad UL_1(i^c - (i - 1)^c)$$

Equate the right-hand sides of the two formulae:

$$UL_1\,(i^c - (i - 1)^c) \quad = \quad UL_1\, c(i + x)^{c-1}$$

Now solve for *x*, with UL_1 dividing out on both sides and with rearrangement:

$$(i + x)^{1-c} \quad = \quad c/(i^c - (i - 1)^c)$$

and:

$$i + x \quad = \quad [c/(i^c - (i - 1)^c)]^{1/(1-c)}$$

so that:

$$x = [c/(i^c - (i-1)^c)]^{1/(1-c)} - i$$

Taking a range of values of c (say, 0.9, 0.8, 0.7, 0.6) and a range of small values of i (say, 2, 4, 6, 8, 10), values of x which may be computed range between –0.540 (for $c = 0.6$ and $i = 2$) and –0.505 (for $c = 0.9$ and $i = 10$), and all the other values of x which may be computed for any value of c as just given in combination with any value of i as just given are likewise near to –0.5.

Berghell concluded that, although the values of x are not constant 'for practical use the variation [in x] is slight and [x] can be considered constant' with the value –0.5 (*Production Engineering in the Aircraft Industry*, p. 187).

Consequently a more accurate approximate formula for small values of i is as follows:

$$UL_i = UL_1 \, c \, (i - 0.5)^{c-1}$$

This second variant is almost the same as the approximate formula but it includes Berghell's constant of –0.5. This formula may be referred to as the 'approximate formula with –0.5'.

We thus have the following five formulae for the particular item unit labour requirement, UL_i:

$UL_i = CTL_{(i)} - CTL_{(i-1)}$

Particular item unit labour requirement learning curve:

$UL_i = UL_1(i^c - (i-1)^c)$ — *exact formula*

$UL_i = UL_1 \, i^c \, (1 - [1 - (1/i)]^c)$ — *simplified exact formula*

$UL_i = UL_1 \, c \, i^{c-1}$ — *approximate* formula*

$UL_i = UL_1 \, c(i - 0.5)^{c-1}$ — *approximate* formula with –0.5 ***

* Strictly speaking, the '=' in this formula should be '≈'.

** To be used whenever i is small.

The first formula is simply a calculation from two successive cumulative total labour requirements and is not a learning curve as such. Nevertheless it can be quite a useful formula, and it shows us how the particular item unit labour requirement may be calculated from the basic cumulative total labour requirement function.

The four formulae following give the particular item unit labour requirement (a measure of performance) as a function of i, cumulative total output, and so may be identified as learning curves, and specifically as the particular item unit labour requirement learning curve.

The exact formula and the simplified exact formula may be applied whenever an especially accurate estimation is called for.

The approximate formula is commonly identified as the learning curve, but we now see that this is simply one among a family of learning curves related by common descent from the cumulative total labour requirement function.

The approximate formula with –0.5 may be used for small values of i, but my preference, for small values of i, would be to use the exact formula.

Table 7.3 gives some examples of estimates of the particular item unit labour requirement for the construction of Liberty ships in Yard F according to the values of $\hat{UL}_1$ and $\hat{c}$ from the cumulative total labour requirement function. As a standard of comparison, in each case we use the estimate of the unit labour requirement from the simplified exact formula.

For $i = 5$, the value of $\hat{UL}_5$ estimated from the simplified exact formula is 1,196,483 labour-hours (the quantity of labour-hours estimated as being required to build the fifth vessel), while the value of the estimate of $\hat{UL}_5$ from the approximate formula with –0.5 (which we should use for small values of i) is 1,195,930 labour-hours, so the difference between the two is very small.

For $i = 10$, the value of $\hat{UL}_{10}$ estimated from the simplified exact formula is 1,039,149 labour-hours (note the decrease from $\hat{UL}_5$); while the value of $\hat{UL}_{10}$ estimated from the approximate formula with –0.5 is 1,039,041 labour-hours, a negligible difference; and the value of $\hat{UL}_{10}$ estimated from the approximate formula is 1,029,059 labour-hours, a very small difference. However, this example does show the benefit of using the approximate formula with –0.5 for small values of i.

For $i = 105$, the value of $\hat{UL}_{105}$ from the simplified exact formula is 661,673 labour-hours, while the value of $\hat{UL}_{105}$ from the approximate formula is 661,078, a negligible difference. This shows that for large values of i the approximate formula provides an accurate estimator of UL_i.

For Yard F, the reported actual batch average unit labour requirement for vessels nos. 101 to 110 inclusive was 652,000 labour-hours, so we can see that the estimated unit labour requirement for vessel no. 105 and the actual batch average are nearly the same.

The batch average unit labour requirement for a batch of any size x can be

Table 7.3
Examples of use of the estimating formulae for unit labour requirement

Foot-note	i	Estimating formula for $\hat{UL}_i$	Estimated unit labour requirement, labour-hours, $\hat{UL}_i$
(1)	5	$1{,}955{,}201 \times 5^{0.8118}[1 - (1 - \frac{1}{5})^{0.8118}]$	1,196,483
(2)	5	$1{,}955{,}201 \times 0.8118\ (5 - 0.5)^{0.8118-1}$	1,195,930
(1)	10	$1{,}955{,}201 \times 10^{0.8118}[1 - (1 - \frac{1}{10})^{0.8118}]$	1,039,149
(2)	10	$1{,}955{,}201 \times 0.8118\ (10 - 0.5)^{0.8118-1}$	1,039,041
(3)	10	$1{,}955{,}201 \times 0.8118\ 10^{0.8118-1}$	1,029,059
(1)	105	$1{,}955{,}201 \times 105^{0.8118}[1 - (1 - \frac{1}{105})^{0.8118}]$	661,673
(3)	105	$1{,}955{,}201 \times 0.8118 \times 105^{0.8118-1}$	661,078

(1) Estimating equation is the simplified exact formula.
(2) Estimating equation is the approximate formula with –0.5.
(3) Estimating equation is the approximate formula.

Source: see text.

estimated from the cumulative total labour requirement function as $(C\hat{TL}_{(i+x)} - C\hat{TL}_{(i)})/x$.

For example, in the case of Yard F the batch average unit labour requirement for a batch of $x = 10$ from $i = 100$ to $i + x = 110$ could have been calculated as:

$$\text{Estimated batch average unit-labour requirement} = \frac{C\hat{TL}_{(110)} - C\hat{TL}_{(100)}}{10}$$

$$= \frac{1{,}955{,}201 \times 110^{0.8118} - 1{,}955{,}201 \times 100^{0.8118}}{10}$$

$$= \frac{88{,}796{,}120.6 - 82{,}184{,}782.6}{10}$$

$$= \frac{6{,}611{,}338}{10}$$

$$= 661{,}133.8 \text{ labour-hours}$$

This estimate differs from the actual recorded batch average by 1.4 per cent.

Estimation is generally more reliable for a cumulative average or for a batch average than for a single item, so it is well to be cautious with any estimate of the unit labour requirement for a particular item and to operate on the assumption of a reasonable chance of, say, an error of *plus or minus* ten per cent.

7.4 Estimating and using the learning curve

Whether we use the cumulative average unit labour requirement, $CAL_{(i)}$, or the unit labour requirement, UL_i, as a measure of performance, in each case the learning curve has the same elasticity, namely $c - 1$, derived from the same cumulative total labour requirement elasticity, c. This is why I emphasize that the cumulative total labour requirement function should be regarded as the basic function from which the learning curves may be derived and that the parameter c should be regarded as the basic parameter of the learning curve.

In general, when estimating the learning curve for any particular set of data it is preferable to estimate the cumulative total labour requirement function. This provides all the parameters required for any learning curve analysis; namely, the parameters UL_1 and c.

We may look at some empirical values of c for thirteen shipyards building Liberty ships during the Second World War. The estimates of c made for these shipyards are given in Table 7.4. (These estimates were obtained by converting the batch average unit labour requirements of Bulletin 824's table 23, pp. 40–41, to batch total labour requirements, thence calculating cumulative total labour requirements, and then applying least-squares to the (natural) logarithms of the variables in the cumulative total labour requirement function. Three of the sixteen shipyards of Table 6.4, Yards N, O, and P, were omitted

Table 7.4
Learning curves for shipyards building Liberty ships

Ship-yard	Elasticity of cumulative total labour requirement w.r.t. cumulative total output, c	Elasticity of cumulative average unit labour requirement w.r.t. cumulative total output,[(a)] $c-1$	Learning curve rate, per cent, $(2^{c-1}) \times 100$
Yard A	0.83797	–0.16203	89.4
Yard B	0.72094	–0.27906	82.4
Yard C	0.76839	–0.23161	85.2
Yard D	0.72217	–0.27783	82.5
Yard E	0.76249	–0.23751	84.8
Yard F	0.81176	–0.18824	87.8
Yard G	0.79740	–0.20260	86.9
Yard H	0.82415	–0.17585	88.5
Yard I	0.79016	–0.20984	86.5
Yard J	0.77130	–0.22870	85.3
Yard K	0.65179	–0.34821	78.6
Yard L	0.65079	–0.34921	78.5
Yard M	0.66213	–0.33787	79.1
Unweighted average	0.751649	–0.24835	84.3

(*a*) The learning curve elasticity.

w.r.t. = with respect to
Source: see text.

because these have observations on only two batches which is insufficient for statistical purposes. The correlation coefficients for the thirteen cumulative total labour requirement functions range between 0.999987 and 0.997083, so in each case the fit is very close and the empirical value of c for each shipyard is well determined.)

Table 7.4 also gives the derived estimates of the learning curve elasticity, $c - 1$, and the learning curve rate expressed as a percentage, $(2^{c-1})100$.

The important thing to observe in Table 7.4 is that, without exception, each shipyard experienced a substantial learning effect. The unweighted average of the cumulative total labour requirement elasticities is 0.75, the unweighted average of the learning curve elasticities is –0.25, and the unweighted average learning curve rate is 84 per cent, meaning that on each occasion that cumulative total output doubled, unit labour requirement (whether cumulative average or particular item) would be reduced by 16 per cent to 84 per cent of its former value.

Table 7.4 demonstrates, first, the substantial impact which the learning effect has, and second, the fact that the learning effect was not an isolated phenomenon in Yard F but was common to all shipyards.

Table 7.4 also shows how comparisons of the learning effect may be made among plants with differing total lengths of production run. The simplest way is to compare the learning curve rate, and we can see that Yard F, the shipyard we used for illustrative purposes, was pretty near to being an 'average' plant because it had a learning curve rate of 87.8 per cent.

We may now proceed to show that the cumulative total labour requirement function may be applied to data reported by period. To do this we may return to the data on the assembling of propelling pencils by an individual worker first seen in Table 6.1. The data of Table 6.1 may be converted to cumulative totals, as shown in Table 7.5. From these data we may make the following estimate of the cumulative total labour requirement function:

$$C\hat{T}L_{(i)} = 0.01924 \times i^{0.8344}$$

with a correlation coefficient of 0.9997.

Again, we see that the cumulative total labour requirement elasticity, *c*, is positive but less than 1 and has a value of 0.8344.

The value of the constant term, 0.01924 is an estimate of the unit labour requirement (in a fraction of a labour-hour) for the first mechanism to be assembled. There are 3,600 labour-seconds in an hour, so we can reckon that the first mechanism took the worker, working alone, about 69.3 seconds to assemble.

The derived value of the learning curve elasticity is –0.1656 (= 0.8344 – 1), and so we can write the particular item unit labour requirement learning curve for this worker as follows (using the approximate formula):

$$\begin{aligned} \hat{UL}_i &= UL_1\, c\, i^{c-1} \\ &= 0.01924 \times 0.8344 \times i^{0.8344-1} \\ &= 0.01605 \times i^{-0.1656} \end{aligned}$$

Thus, despite the presentation by periods of the data in Table 6.1, we can, using the cumulative total labour requirement function, derive the particular item unit labour requirement learning curve for this worker.

Table 7.5
Estimating the cumulative total labour requirement function for the assembling of propelling pencil mechanisms

Cumulative total of mechanisms assembled, i	Cumulative total labour requirement, labour-hours used, $CTL_{(i)}$
9,300	40
21,200	80
35,800	120
51,100	160
66,900	200
82,800	240
99,000	280
115,000	320
131,300	360
147,700	400
163,700	440
180,700	480

Source: Table 6.1.

The impact of the learning effect may be demonstrated by using the particular item unit labour requirement learning curve. This worker started by assembling the first propelling pencil mechanism in 69.3 seconds. By the time this worker had assembled 180,700 mechanisms, the particular item unit labour requirement would have been (using the approximate formula):

$$\hat{UL}_{180,700} = 0.01924 \times 0.8344 \times 180{,}700^{-0.1656}$$

$$= 0.00216 \text{ labour-hours}$$

or, in the case of this worker working alone, 7.8 seconds!

For the worker assembling propelling pencil mechanisms, the derived value of the learning curve rate is 89 per cent (= $(2^{-0.1656})100$). Thus we can see that this individual worker assembling propelling pencil mechanisms had a learning curve rate which was not all that different from the learning curve rate in an average shipyard building Liberty ships (84 per cent – Table 7.4). This helps to show the general applicability of the learning curve analysis to any situation where learning is occurring.

7.5 Unit cost and the learning curve

In order to estimate unit cost using the learning curve, a transition has to be made from the behaviour of cumulative total labour requirement (and the consequential unit labour requirement) to the behaviour of cumulative dollar cost (and the consequential unit cost).

This transition is more complex than is commonly supposed. The common procedure is simply to assume, or to postulate, that a curve of exactly the same form as the learning curve applies to the cost of production. Under what conditions would this procedure work?

The transition from the learning curve for unit labour requirement to the curve for the behaviour of unit cost should begin with unit labour cost. We have seen that the particular item unit labour requirement follows a learning curve (approximate formula) of the form $UL_i = UL_1 \, c \, i^{c-1}$, where the derivation is ultimately from the cumulative total labour requirement function.

Where p_L is the price for a unit of labour, unit labour cost for the ith item is simply $p_L \times UL_i$. Thus we may derive a unit labour cost for the ith item as:

$$\text{Unit labour cost for } i\text{th item} = p_L \times UL_i$$

$$= (p_L \times UL_1) \, c \, i^{c-1}$$

Because $(p_L \times UL_1)$ is the unit labour cost for the first item to be produced, we may by substitution write:

$$\text{Unit labour cost for } i\text{th item} = \begin{pmatrix}\text{Unit labour}\\ \text{cost for}\\ \text{first item}\end{pmatrix} c \, i^{c-1}$$

This is a learning curve, because unit labour cost can be taken as a measure of performance, which is here given as a function of cumulative total output, i.

The unit labour cost curve for the ith item thus has the same elasticity with respect to i as does the particular item unit labour requirement learning curve.

Next, we need to consider unit materials cost. The analysis initially assumes the input of only one material. Extending the analysis to more than one material input is straightforward and will be briefly described subsequently.

Suppose that unit material cost decreases as cumulative total output increases. In his 1936 article (already referred to), Wright (who had first-hand knowledge of these matters over fifteen years of aircraft production at the Curtiss-Wright Corporation) observed: 'Material also decreases in cost as quantity increases' (p. 125).

Wright gave two reasons why this happened. In the first place, 'the amount of waste is cut down ... in a quantity of one to five units [airplanes] this waste is as large as forty percent. It reduces rapidly as quantity increases, to twenty-five or thirty percent in quantities of twenty-five to fifty units and down to twenty percent in a quantity of one hundred units.' Wright ascribed this reduction in waste of materials to greater efficiency resulting from learning and experience in working with materials. This is something that occurs directly in the given plant.

But this was not all. In the second place, Wright observed there was the reduction in unit cost occurring in the supplier's plant (also due to learning) and this would affect (reduce) the price charged by the supplier. Additionally there was commercially the possibility of negotiating 'greater discounts [for materials supplied] when purchasing in large quantities'.

If it is only (physical) unit material requirement in the given plant which benefits from the learning effect, then we can define a cumulative total material requirement function and proceed as for the derivation of the learning curve relating to labour input. Let $CTM_{(i)}$ denote cumulative total material requirement and UM_i denote unit material requirement for the ith item. Furthermore, we denote the cumulative total material requirement elasticity as c' (with the restriction that $c' < 1$). Then:

$$CTM_{(i)} = UM_1 \, i^{c'}$$

From which may be derived (by the same route as for the particular item unit labour requirement learning curve (approximate formula)):

$$UM_i = UM_1 \, c' \, i^{c'-1}$$

If the price per unit of material, denoted p_M, is constant, then, letting UMC_i denote the unit material cost for the ith item:

$$\begin{aligned} UMC_i &= p_M \, UM_1 \, c' \, i^{c'-1} \\ &= UMC_1 \, c' \, i^{c'-1} \end{aligned}$$

Under the condition that only the unit material requirement in the given plant is subject to the learning effect and the material price is constant, we need to assume that the cumulative total material requirement elasticity has the same value as the cumulative total labour requirement elasticity, that is, $c' = c$. This I call the 'parameter assumption (constant price)'. The purpose of the parameter assumption (constant price) is to permit subsequent factorization in the derivation of the unit cost learning curve. Using the parameter assumption (constant price):

$$UMC_i \quad = \quad UMC_1\, c\, i^{c-1}$$

If the condition is that both unit material requirement and also the price for material are subject to a learning effect, then things become more complex.

Let p_{Mi} denote the now-varying (decreasing) price charged by the supplier for the material used in the ith item. We can postulate the following function for p_{Mi}, where p_{Mi} decreases with cumulative total output, with the restriction that i does not equal 1 in the postulated function:

$$p_{Mi} \quad = \quad p_{M1}\, d\, i^{c'-1}$$

where d is a constant whose value will be shortly explained and c' is the cumulative total material requirement elasticity.

By definition:

$$UMC_i \quad = \quad p_{Mi}\, UM_i$$

Whence by substitution:

$$\begin{aligned} UMC_i \quad &= \quad p_{M1}\, d\, i^{c'-1}\, UM_1\, c'\, i^{c'-1} \\ &= \quad p_{M1}\, UM_1\, d\, c'\, i^{2c'-2} \end{aligned}$$

We now need to assume that, where c is the cumulative total labour requirement elasticity:

$$d \quad = \quad 2c/(c+1)$$

and:

$$c' \quad = \quad (c+1)/2$$

These two equations I call the 'parameter assumptions (decreasing price)'. The purpose of the parameter assumptions (decreasing price) is to permit subsequent factorization in the derivation of a unit cost learning curve. In order to derive the unit cost learning curve we have either to make the parameter assumption (constant price) or the parameter assumptions (decreasing price);

when I wish to refer to both of these parameter assumptions, as appropriate to the conditions, I shall simply refer to the 'parameter assumptions'.

With the parameter assumptions (decreasing price):

$$d\,c' \quad = \quad \frac{2c}{c+1} \times \frac{c+1}{2} \quad = \quad c$$

and:

$$2c' - 2 \quad = \quad 2 \times \frac{c+1}{2} - 2 \quad = \quad c - 1$$

So that, under the parameter assumptions (decreasing price):

$$\begin{aligned} UMC_i \quad &= \quad p_{M1}\, UM_1\, c\, i^{c-1} \\ &= \quad UMC_1\, c\, i^{c-1} \end{aligned}$$

If there is more than one (type of) material used as an input to production, then this analysis can be repeated for each type of material (repeating the parameter assumption (constant price) or the parameter assumptions (decreasing price) as appropriate).

The assumption that $c' = (c + 1)/2$, together with the condition $0 < c < 1$, means that c' is greater than c (as follows: $1 > c$; therefore $1/2 > c/2$; therefore $c/2 + 1/2 > c/2 + c/2$ or $c' > c$). In other words, when the material price also decreases with cumulative total output, then, for the derivation of the unit cost learning curve, unit material requirement should be less subject to the learning effect than is unit labour requirement.

The result that $c' > c$ allows for the additional impact on unit material cost of the declining price for material. A greater value of c' than c would seem plausible, because we can readily envisage how labour becomes more efficient at tasks, whereas it is less easy to envisage materials being economized in the same systematic way, or at least to the same extent.

The question then arises as to the behaviour of unit capital cost. This is a complicated issue because unit capital requirement must be calculated as the acquisition cost of capital installed divided by the *annual* quantity of output produced. (This is necessary to conform with the resource price of capital being taken as the annual rate of depreciation.) In general, textbooks fail to consider the impact of the learning effect on the time rate of production; we have not so far discussed the impact of the learning effect on the time rate of production but this will be done fully in the next chapter.

Consequently, it is common to sidestep the matter of unit capital cost simply by not incorporating unit capital cost into the analysis. Accordingly, the conventional analysis simply proceeds in terms of unit material and labour cost (combined), and this procedure is usually more implicit than explicit. In other words, the conventional analysis is confined to the *cash outflow* resource costs of production for labour and for materials. These costs I call the 'cash cost' of production.

If we thus ignore unit capital cost, we may write, either for the condition that only unit material requirement is subject to the learning effect, when we need to assume $c' = c$; or for the condition that both unit material requirement and material price are subject to a learning effect, when we need to assume $d = 2c/(c + 1)$ and $c' = (c + 1)/2$:

$$\begin{array}{l} \text{Unit cash} \\ \text{cost for} \\ i\text{th item} \end{array} = \begin{array}{l} \text{Unit material} \\ \text{cost for} \\ i\text{th item} \end{array} + \begin{array}{l} \text{Unit labour} \\ \text{labour cost} \\ \text{for } i\text{th item} \end{array}$$

$$= \begin{pmatrix} \text{Unit material} \\ \text{cost for} \\ \text{first item} \end{pmatrix} c\, i^{c-1} + \begin{pmatrix} \text{Unit labour} \\ \text{cost for} \\ \text{first item} \end{pmatrix} c\, i^{c-1}$$

If there is more than one type of material to be included in the analysis, then additional terms in unit material cost can simply be included as requisite (but with the same parameter assumptions as appropriate).

We then have by factorization (and the purpose of the parameter assumptions is to make this factorization possible):

$$\begin{array}{l} \text{Unit cash} \\ \text{cost for} \\ i\text{th item} \end{array} = \begin{pmatrix} \text{Unit material} & & \text{Unit labour} \\ \text{cost for} & + & \text{cost for} \\ \text{first item} & & \text{first item} \end{pmatrix} c\, i^{c-1}$$

$$= \begin{pmatrix} \text{Unit cash} \\ \text{cost for} \\ \text{first item} \end{pmatrix} c\, i^{c-1}$$

This function may be called the 'unit cash cost learning curve', because unit cash cost is a measure of performance, and is here given as a function of cumulative total output, i. When it appears in this function, we shall refer to

$c - 1$ as the 'unit cash cost elasticity'.

The parameters of this function may be estimated by relating cumulative total cash cost to cumulative total output, in exactly the same way as may be done for cumulative total labour requirement.

That is, estimate:

$$\begin{pmatrix}\text{Cumulative}\\ \text{total cash}\\ \text{cost}\end{pmatrix}_{(i)} = \begin{pmatrix}\text{Unit cash}\\ \text{cost for}\\ \text{first item}\end{pmatrix} i^{c}$$

In the cumulative total cash cost function, the elasticity, c, may be referred to as the 'cumulative total cash cost elasticity'. An estimate of the unit cash cost for the first item to be produced in the given plant will also be obtained.

If the parameter assumptions, which we need to make in order to derive by factorization the unit cash cost learning curve, do not hold, which is likely to be the case, then it is still possible to have a unit cash cost learning curve of a sort, but the following two related problems arise, the first less serious, the second more serious in principle.

First, the cumulative total cash cost elasticity, which may be estimated for any given sample of observations on cumulative total output and cumulative total cash cost, is a rather complicated average of the values of the cumulative total elasticities for each of the separate inputs. This means that it is difficult to give any straightforward interpretation to such a cumulative total cash cost elasticity, other than noting its 'average' nature.

Second and relatedly, the estimate of the elasticity will depend upon the length of the production run over which the value for the parameter is estimated.

That is, if we have data (a sample of observations by period or by batches on cumulative total cash cost and cumulative total output) for a production run totalling j items, then we may estimate from this sample of observations a value of c which may be denoted as $c\{j\}$ where '$c\{j\}$' is to be read as: estimate of c with respect to cumulative total j.

If we then take a production run of $2j$ items with another (increased) sample of observations on cumulative total cash cost and cumulative total output, then we obtain $c\{2j\}$.

The problem that arises (when there are differing values of c among the different inputs) is that $c\{j\}$ will differ from $c\{2j\}$, and we would find that $c\{j\} < c\{2j\}$. That is, the longer the production run from which the sample of observations is taken, the greater the estimate of c.

This happens because those inputs with the greater cumulative total elasticity (in other words, those inputs which are less responsive to the learning effect) systematically become a greater proportion of total cash cost. This is because the unit input requirements of the inputs with the greater values of c do not decline as rapidly as the unit input requirements of those inputs with smaller values of c. Consequently, the 'weighting' of such (less responsive) elasticities in the estimation of the 'average' elasticity becomes greater. Consequently, in turn, the estimate of the 'average' elasticity increases.

Eventually, the 'average' elasticity approaches the value of the greatest elasticity c among the cumulative total input elasticities.

Another way of looking at this second problem is that, because of differing elasticities of cumulative total input requirement with respect to i among the different inputs, the behaviour of the cumulative total cash cost function, and hence the behaviour of the unit cash cost learning curve, is each no longer linear in logarithms. Plotted on double-logarithmic graph paper, the plot points for cumulative total cash cost against i, or for unit cash cost for the ith item against i, will each form a curve rather than a straight line.

This is in principle a serious problem for the extended learning curve analysis, because one of the virtues of the learning curve, as so far discussed, is that the estimate of the cumulative total elasticity is in principle independent of the length of the production run from which any sample of observations is taken. In other words, so far the estimating procedure has been based on a function linear in logarithms.

In practice, it may turn out that the differences in the estimates of the cumulative total cash cost elasticity for production runs of differing length are quantitatively only small. It is, effectively, the possibly empirically small nature of the second problem which makes it permissible to use the unit cash cost learning curve even when the elasticities differ among inputs. But even so, cost estimation in these circumstances will be subject to a greater margin of error than would be the case when the parameter assumptions hold.

Nevertheless, the cumulative total cash cost function and the unit cash cost learning curve are useful tools in cost estimating, and are widely used as such.

Having explained the learning curve in all its various forms, we now need to consider the consequences of the learning effect and the learning curve for the (elapsed) time rate of production. What happens to output per period under the learning effect? What happens to period average unit cost under the learning effect? What happens to the rate of profit under the learning effect?

These questions will be answered in the next chapter.

8 Learning and Profitability

8.1 Learning and the time rate of production

An important consequence of the learning effect is that a plant which uses a given and constant quantity of total labour-hours per period will find that the quantity of physical output it produces per period will increase from one period to the next.

For example, if a plant consists of 3 workers, each of whom works a 40-hour week, its total labour input will be 120 labour-hours per week. If at first the plant's period average unit labour requirement is 10 labour-hours per item, the plant will produce at a time rate of 12 items per week (= 120 labour-hours per week/10 labour-hours per item). (Throughout this chapter, the unqualified word 'time' means 'elapsed time', so that 'time rate of production' means 'rate of production per period of elapsed time'.)

If the plant's period average unit labour requirement decreases to 8 labour-hours per item (because of the learning effect) and if the plant's total labour input remains 120 labour-hours per week, then the plant's time rate of production will increase to 15 items per week (= 120 labour-hours per week/ 8 labour-hours per item).

As the period average unit labour requirement continues to decrease, because of the continuing impact of the learning effect on the particular item unit labour requirement, the plant's time rate of production will continue to increase (providing the total labour input per period remains constant).

This is the consequence we must now explore because it is an important economic consequence of the learning effect.

The learning effect also makes it possible for a plant to maintain a constant time rate of production while reducing the total labour input, either by reducing the number of workers or by reducing the hours worked per worker per period (the latter being more likely), but we shall not consider this alternative of a reduction in total labour input per period, and throughout we assume a constant input of labour per period.

Let us illustrate how this might have affected Yard F building Liberty ships, taking each of vessels 1 to 20 inclusive. The estimated cumulative total labour requirement function for Yard F was (see Chapter 7):

$$C\hat{T}L_{(i)} = 1{,}955{,}201 \times i^{0.8118}$$

Inserting for i each in turn of the values 1, 2, 3, ..., 20, we may estimate the required series on $C\hat{T}L_{(i)}$. The results are given in the first column of data in Table 8.1.

From the estimates of cumulative total labour requirement, the unit labour requirement for each vessel may readily be calculated by subtraction, according to the equation $U\hat{L}_i = C\hat{T}L_{(i)} - C\hat{T}L_{(i-1)}$. The results are given in the second column of Table 8.1.

In the second column of Table 8.1 we see the feature of a decreasing unit labour requirement, consequent upon the learning effect. We note how this decrease is especially rapid at first. (With a constant price per labour-hour this means that the unit labour cost will decrease also, and this is a feature which is of great importance to cost accountants.)

How would this decrease in unit labour requirement have affected the time rate of production in Yard F? We could answer this question in actuality if we knew the total labour input per period of time which Yard F used, but this information is not available. However, we may give an hypothetical illustrative answer if we assume some constant and specific labour input used by Yard F. For this we have the information that the average size of the labour force in private shipyards building Liberty ships was 3,056 workers, and the length of the working week was, on average, 47.7 hours (United States Department of Labor Bureau of Labor Statistics, *Wartime Employment, Production, and Conditions of Work in Shipyards*, Bulletin No. 824, 1945, hereafter referred to as Bulletin 824, table 5, p. 7 and p. 17). This makes the total input of labour, in the average yard, 145,771.2 labour-hours per week.

Let us suppose that Yard F used a total labour input of 145,771.2 labour-hours per week and that this remained constant. Given that Yard F built 173 Liberty ships in about three years, it is certain that Yard F had more than one building berth and a labour force considerably greater than 3,056 workers, but this does not affect the principle of the illustration, which we can take as applying simply to one berth (or shipway or plant) and its associated labour force. We assume that a berth in Yard F completed the construction of each vessel before starting upon construction of the next vessel in sequence.

Taking this hypothetical situation, we can estimate (hypothetically) the elapsed time, in weeks, for each of the vessels built by a berth in Yard F simply by dividing the assumed (and constant) input of total labour-hours per

Table 8.1
Estimated unit labour requirement and consequent hypothetical elapsed time per vessel in Yard F

Vessel no., i	Estimated labour-hours: Cumulative total labour requirement,[a] $\hat{CTL}_{(i)}$	Estimated labour-hours: Unit labour requirement for ith vessel,[b] $\hat{UL}_i$	Hypothetical estimated elapsed time per vessel,[c] weeks, $\hat{E}_i$
1	1,955,201	1,955,201	13.4
2	3,432,071	1,476,870	10.1
3	4,769,804	1,337,733	9.2
4	6,024,502	1,254,698	8.6
5	7,220,863	1,196,361	8.2
6	8,372,699	1,151,836	7.9
7	9,488,780	1,116,081	7.7
8	10,575,139	1,086,359	7.5
9	11,636,163	1,061,024	7.3
10	12,675,176	1,039,013	7.1
11	13,694,779	1,019,603	7.0
12	14,697,057	1,002,278	6.9
13	15,683,716	986,659	6.8
14	16,656,175	972,459	6.7
15	17,615,635	959,460	6.6
16	18,563,121	947,486	6.5
17	19,499,516	936,395	6.4
18	20,425,594	926,078	6.4
19	21,342,032	916,438	6.3
20	22,249,431	907,399	6.2

(*a*) Estimated from cumulative total labour requirement function:

$$\hat{CTL}_{(i)} = 1{,}955{,}200.8 \times i^{\,0.81176264}$$

(*b*) Estimated from equation: $\hat{UL}_i = \hat{CTL}_{(i)} - \hat{CTL}_{(i-1)}$ using estimated values of $CTL_{(i)}$.

(*c*) The average private shipyard had 3,056 workers, each working an average of 47.7 hours per week, making a total labour input of 145,771.2 labour-hours per week in the average shipyard. Assuming that this labour input can be applied to Yard F, and assuming also that Yard F had only one berth and so completed the building of each vessel before starting on construction of the next, we can estimate an hypothetical elapsed time in weeks per vessel, $\hat{E}_i$, by dividing estimated unit labour requirement $\hat{UL}_i$ by 145,771.2 labour-hours per week.

Source: see text.

week into the labour requirement for each vessel.

The mathematics of this estimation is straightforward. Let:

E_i = The number of weeks (periods) of elapsed time required to build vessel *i*

N = The number (stock) of workers employed, assumed constant

H = The average number of hours worked per worker per week (period), assumed constant

Then the labour-hour input required to build vessel *i*, UL_i is:

$$UL_i = E_i \times N \times H$$

Whence:

$$E_i = UL_i/(N \times H)$$

We have assumed a value of N (3,056 workers) and a value of H (47.7 hours per week) and we have also assumed that both N and H remain constant. Thus, using the estimated values of $\hat{UL}_i$ for Yard F we may estimate a set of hypothetical values for $\hat{E}_i$. The results are given in the third column of Table 8.1.

These values are hypothetical but are nevertheless very informative as to the sort of thing which happened to the elapsed time required for the production of Liberty ships as each shipyard proceeded down its learning curve.

The first vessel produced in Yard F would have taken 13.4 weeks of elapsed time to construct; the second vessel 10.1 weeks of elapsed time; the third vessel 9.2 weeks of elapsed time; and so on, until the twentieth vessel would have taken 6.2 weeks (!) of elapsed time.

We can try to show in a more conventional way what would have happened to the time rate of production in a berth in Yard F. We may observe that vessel no. 1 and vessel no. 2 would together have taken 23.5 weeks of elapsed time to construct. This is near enough to being half a year, so we can say that in the first half-year of production a berth in Yard F would have constructed (approximately) two Liberty ships.

We may then observe that vessels nos. 3 to 5 inclusive would together have taken 26 weeks of elapsed time to construct; this is exactly half a year, so we can say that in the second half-year of production a berth in Yard F would have constructed three Liberty ships.

Proceeding in this way we may derive the approximate figures for the rate of production per half-year, as shown in Table 8.2. You can see that in the fourth half-year the time rate of production has increased to four Liberty ships per half-year. In other words, the time rate of production has doubled by comparison with the first half-year!

These hypothetical estimates of the time rate of production are in line with recorded data on the time rate of production of Liberty ships of 4 vessels per shipway per annum, increasing to 8 vessels per shipway per annum (Frederic C. Lane, *Ships For Victory A History of Shipbuilding under the U.S. Maritime Commission in World War II*, pp. 139–40, 174–6 and 210; Bulletin 824, pp. 42–3).

Had Hitler known about the particular item unit labour requirement learning curve and its consequences for the time rate of production he might not have been quite so confident about winning the Battle of the Atlantic! With this evidence, we should not be surprised that the United States shipbuilders won the vital wartime race, and indeed produced in two years a greater tonnage than had been directed by President Roosevelt in February 1942 – a directive that had, at the time, been widely regarded as impossible to accomplish.

Finally in this section, we may show the probable course of the time rate of production per day in the case of the worker assembling propelling pencil mechanisms, taking data for two weeks only; this also shows how Figure 6.4 was derived.

Using the cumulative total labour requirement function we may derive estimates of cumulative total output produced, $\hat{i}$, at each of the junctures where the cumulative total labour requirement in labour-hours came to the

Table 8.2

Hypothetical, and approximate, time rate of production in Yard F

Half-year	Vessels built, nos.	Time rate of production, vessels built per half-year
First	1 to 2 inclusive	2
Second	3 to 5 inclusive	3
Third	6 to 8 inclusive	3
Fourth	9 to 12 inclusive	4
Fifth	13 to 16 inclusive	4
Sixth	17 to 20 inclusive	4

Source: Table 8.1.

end of each 8-hour day. Table 8.3 gives for each of cumulative total labour requirement, $CTL_{(i)}$, at 8, 16, 24, ..., 80 labour-hours, the corresponding estimate of the cumulative total output, $\hat{i}$, according to a rearrangement of the cumulative total labour requirement function.

These estimates of $\hat{i}$ are given in the second column of data in Table 8.3. By subtracting each preceding figure on $\hat{i}$ we obtain, in the third column, an estimate of the total number of mechanisms assembled during that particular 8-hour day. Finally, we may divide each of these daily outputs into 8 labour-

Table 8.3
Estimating the impact of decreasing unit labour requirement on the time rate of assembling propelling pencil mechanisms per day[a]
(first two weeks of production)

Day	Cumulative total labour requirement, $CTL_{(i)}$	Estimated[a] cumulative total of mechanisms assembled, $\hat{i}$	Number of mechanisms assembled that day	Performance during that day: labour-hours per thousand mechanisms
1	8	1,375	1,375	5.818
2	16	3,156	1,781	4.492
3	24	5,131	1,975	4.051
4	32	7,243	2,112	3.788
5	40	9,464	2,221	3.602
6	48	11,775	2,311	3.462
7	56	14,165	2,390	3.347
8	64	16,623	2,458	3.255
9	72	19,143	2,520	3.175
10	80	21,719	2,576	3.106

(*a*) Estimated from the empirical cumulative total labour requirement function (see Table 7.5 and using natural logarithms):

$$\hat{CTL}_{(i)} = 0.019242447 \times i^{0.834439681}$$

and by rearrangement but using actual $CTL_{(i)}$ at the end of each day:

$$\hat{i} = \left(\frac{CTL_{(i)}}{0.019242447}\right)^{1/0.834439681}$$

Source: see texts of Chapters 7 and 8 and Table 7.5.

hours in order to obtain, in the last column, the average unit labour requirement during that day (expressing the result in labour-hours per thousand mechanisms for convenience).

We can then see two things. First, the time rate of output per day increases under the impact of the learning effect. Again, the increase is rapid in the first few days but the rate of increase gradually slows with each passing day (period).

Second, and consequently, the day average performance of the worker, as measured inversely in terms of unit labour requirement, improves quickly at first, and subsequently more slowly. Or, to put the matter more exactly, it is the improving performance of the worker, in conjunction with the constant labour input of 8 labour-hours per day, which leads to the increasing time rate of output per day.

These values of $\hat{i}$ were used for Figure 6.4, where points C to L inclusive plot the last column of Table 8.3 against the second column. But the really important point about Table 8.3 is to illustrate once again how the learning effect leads to an increased time rate of output.

In the top part of Figure 6.2, we saw the pattern of improvement from Week 0 to Week 11 in the weekly output produced by the worker assembling propelling pencil mechanisms: weekly output increased rapidly at first and then the improvement slowed. This is a typical pattern of improvement under the learning effect, which tends to have its greatest impact in the first few periods.

We now explain how these important changes in the time rate of output may be described and analysed by using the time constant progress function.

8.2 The time constant progress function

A general function has been derived which models or describes the trend of output per period (or, alternatively, of period average productivity) in a plant which is subject to the learning effect. If the labour input per period is constant, then the increase in period average labour productivity will be proportionately the same as the increase in output per period. For example, if the assembling of propelling pencil mechanisms were taking place in a plant with two workers, then period average labour productivity (output per worker per week) would rise proportionately as in the second column of Table 6.1 and output per week from the plant would simply be each period's average productivity multiplied by two. Because we are assuming constant labour input, further discussion will be in terms of the time rate of production, or the quantity of output produced per period of elapsed time, but productivity will be mentioned in brackets as a reminder that the argument is generally applicable.

The general function which describes the pattern of change seen in the top part of Figure 6.2 may be called the 'time constant progress function'.

I refer to this function as a 'progress function' because it gives performance as a function of the elapse of time. (However, most of the literature uses confusing terminology and refers to the function as the 'time constant learning curve'.)

The function is specifically identified by its key parameter, the time constant. The time constant is the most important among the three parameters of this function. (The first published paper specifically proposing the time constant progress function (in a model analogous with the function used in physics) appears to be F. W. Bevis, C. Finniear, and D.R. Towill, 'Prediction of operator performance during learning of repetitive tasks', *The International Journal of Production Research*, 1970, vol. 8, no. 4, pp. 293–305 (the time constant function seems to have first been used in this context by Bevis in his 1970 MSc Thesis at the University of Wales Institute of Science and Technology); a bibliography of progress function and learning curve literature covering the period 1970 to 1980 is available in a publication of the Institution of Electronic and Radio Engineers, *Industrial Applications of Learning Curves and Progress Functions*, Conference Proceedings No. 52 (IERE, London, 1981), pp. 7–15.)

The time constant is a parameter which measures, for any particular set of data, a constant number of discrete time periods, each of equal duration, which have to elapse before that period in which the plant achieves 63.2 per cent of the difference between initial output per period (or initial period average productivity) and (eventual) maximum output per period (or maximum period average productivity).

Furthermore, after the elapse of 3 times the time constant number of periods, the plant achieves 95 per cent of the difference between initial output per period (or initial period average productivity) and (eventual) maximum output per period (or maximum period average productivity), and so this number of periods can be taken as the number of periods which have to elapse before that period in which the plant will achieve pretty well maximum output per period (or maximum period average productivity).

The reason for these proportions of 0.632, after the elapse of the time constant number of periods, and 0.95, after the elapse of 3 times the time constant number of periods, will be explained shortly.

The time constant parameter can be conceptually compared with the measure devised by the physicist Ernest Rutherford of the radioactive 'half-life' of any radioactive substance (or 'half-value period' as he called it – *Nature*, October 6, 1910, p. 431, based on his research going back to 1902). Radioactive 'half-life' is defined as the number of periods taken for the rate of radioactive emission by a pure sample of the substance to decrease to one-half of the initial rate. The half-life parameter refers to (a rate of) decay or decrease, whereas the time constant usually refers to (a rate of) growth or increase.

Radioactive emissions from a radioactive substance only asymptotically approach zero, so there is no sense to the concept of a total lifetime of a radioactive substance. But it is possible to measure the time in which radioactive emissions will decline by some fraction, and for convenience Rutherford took the fraction one-half. Given the physical characteristics of any radioactive substance, Rutherford's radioactive half-life parameter provides a *unique single measure* whereby to characterize that substance. Thus in textbooks on physics there are comparative tables giving the radioactive half-life of, for example, radium of atomic weight 226 (1,620 years), or the medically and industrially useful cobalt-60 (5.3 years), or thorium X (3.64 days) – the substance originally studied by Rutherford in 1902.

In a similar manner, in any physical system where a physical quantity varies as a function of time and where after some delay the physical quantity asymptotically approaches some maximum, the time constant is a parameter which describes this delay. The time constant thus provides a convenient and unique single measure for characterizing the system and comparing it with other comparable systems.

In the learning effect, we have a 'learning system' in which the initial time rate of production is increased through learning and in which a certain maximum performance is gradually and asymptotically approached.

Accordingly, we can uniquely characterize such a learning system by the time taken in periods of elapsed time before a certain increased rate of production per period can be achieved (that is, for a certain amount of learning to occur). This certain increased rate is taken at 63.2 per cent of the transition from initial output per period (or initial period average productivity) to eventual maximum achievable rate of production per period (or maximum period average productivity). The time taken in elapsed periods before this 63.2 per cent transition is achieved is called the 'time constant' of the learning system. We can then use this single measure of the time constant to characterize and compare the impact of the learning effect in one situation (plant) with the learning effect in another comparable situation (plant).

For example, we could say that in one plant the time constant is 2 periods while in another plant the time constant is 5 periods. That is to say, starting from the initial period of production, one plant will achieve 63.2 per cent of the transition to maximum production per period after the elapse of 2 periods, but the other plant will achieve 63.2 per cent of the transition to maximum production per period after the elapse of 5 periods.

The periods may be hours, days, weeks, months, quarters, half-years, or years, depending on the nature of the industrial processes involved. Obviously, assembling the internal mechanisms of propelling pencils by one worker has a different time-scale of learning than does the building of Liberty ships by a work group of thousands of workers. Because the elapsed time duration of the period involved has to be selected to suit the nature of the process, the time constant can meaningfully be compared only among similar or comparable plants producing the same type of product by the same type of process with the same type of equipment. That is, the same duration of period of production has to be appropriate to each of the plants being considered. Similar remarks apply if performance is being analysed by task. In this respect, the time constant differs from the learning curve elasticity which is *theoretically* comparable among plants or tasks of differing types.

A greater value of the time constant parameter for a learning system tells us that the learning process is slower because it takes longer (in terms of elapsed time) to achieve the 63.2 per cent transition from initial production to eventual maximum production.

The time constant progress function then expresses the rate of production per period during any period, t, as being equal to the sum of two components (the time constant progress function may also express the period average productivity during any period t as being equal to the sum of two components).

The first component of the rate of production is the rate of production during the very first, or 'initial', period of operation. This we shall call the 'initial output' (or 'initial productivity', if productivity is being discussed). The initial output will be the time rate of production during the first period of elapsed time (the initial productivity will be the period average productivity during the first period of elapsed time). The initial output (or initial productivity) is a constant for any particular set of data, and so may be called a parameter of the time constant progress function.

The second component of the rate of production varies with the number of periods which have elapsed and is effectively a varying fraction of the difference between the (eventual) maximum rate of output (or productivity) and the initial output (or initial productivity). This varying fraction may take the value zero (despite this we shall refer to it as a 'fraction'), but otherwise must be positive, and it asymptotically approaches but never reaches the value of unity.

For any particular set of data, the difference between eventual maximum output and the initial output will be a constant, and so may also be called a parameter of the time constant progress function.

In this way, the time constant progress function has three basic parameters or constants: the time constant itself; the initial output (or initial productivity); and the eventual maximum output (or eventual maximum productivity). The difference between the eventual maximum output (or eventual maximum productivity) and the initial output (or initial productivity) is a fourth, derived, parameter of the time constant progress function.

We now discuss the varying fraction which is, so to speak, the working part of the time constant progress function.

In the time constant progress function, the varying fraction varies positively with the number of periods which have elapsed since the beginning of the process. Furthermore, the size of the fraction, for any given period t, depends upon the time constant.

To take this explanation further we use the following terminology and notation (we shall discuss the system in terms of the time rate of production only, but the time constant progress function can also apply to period average labour productivity).

Q_t = The rate (flow) of the physical quantity of output during period t; where t is a specified and definite period of elapsed time; in other words, the time rate of production for t; a variable of the system

Q_0 = Initial output; the rate (flow) of the physical quantity of output during the initial or first specified and definite period of elapsed time; a parameter of the system

Q_{max} = The eventual maximum output; the rate (flow) of the physical quantity of output per specified and definite period of elapsed time; this time rate of production is approached asymptotically; a parameter of the system

t = Discrete and specified and definite periods of elapsed time, all of equal duration, where $t = 0, 1, 2, 3, ..., T, ...$; note that we must use 0 to denote the initial period of time (not 1)

T = The time constant; a parameter of the system, measured as a number of discrete periods of elapsed time, where:

$$\frac{Q_T - Q_0}{Q_{max} - Q_0} = 0.63212$$

= Proportion of transition from Q_0 to Q_{max}

T must be positive and may not be equal to 0

e = The base for natural logarithms: 2.71828...

Note the numbering of the periods starts with the first, or initial, period being numbered 0; subsequent numbers for t then also indicate the number of complete periods, all of equal duration, which have elapsed after the initial period. The value of the time constant T indicates the number of complete periods which have elapsed so that by, or during, $t = T$, the relevant 63.2 per cent transition to maximum output will have been attained.

The explanation of the time constant progress function may begin with the following general equation:

$$Q_t = Q_0 + (Q_{max} - Q_0)\begin{pmatrix}\text{Fraction varying} \\ \text{with } t \text{ and} \\ \text{dependent upon } T\end{pmatrix}$$

If the fraction varying with t takes the value 0, then we have:

$$Q_t = Q_0$$

If the fraction varying with t so nearly approaches 1 as not to be practically different from 1, then we have:

$$Q_t \approx Q_0 + (Q_{max} - Q_0)$$

$$\approx Q_{max}$$

Thus if the varying fraction varies between 0 and the asymptotic limit of 1, then Q_t will vary between Q_0 and asymptotically approaching the limiting value of Q_{max}.

In order to represent the time pattern of increase in output per period, the varying fraction must meet the following five conditions:

(1) when $t = 0$, the value of the varying fraction must also equal zero, so that $Q_t = Q_0$;

(2) the value of the varying fraction must vary positively with t for any given value of the time constant T;

(3) the value of the varying fraction must, as t increases (and for any given value of T), increase quickly at first and then more slowly, so reproducing the pattern shown in the top part of Figure 6.2;

(4) the value of the varying fraction must asymptotically approach a limiting value of 1, so that Q_t will asymptotically approach a limit of Q_{max};

(5) the value of the varying fraction, for any given t, should depend in an appropriate way on the value of the time constant T; that is, for any given t, the value of the varying fraction should be lower the greater the value of T.

A function which will give a varying fraction of the required sort is the following:

$$\begin{pmatrix}\text{Fraction varying}\\ \text{with } t \text{ and}\\ \text{dependent upon } T\end{pmatrix} = 1 - \frac{1}{e^{t/T}}$$

$$= 1 - e^{-t/T}$$

$$= 1 - 2.71828^{-t/T}$$

The second line of the equation gives the form in which the varying fraction is most commonly expressed and is the form we shall use later on, but the first line of the equation is also of use because it shows how the denominator increases, and at first we shall mainly use the form shown in the first line.

Except for the case where $t = 0$, t and T are both positive, and because e has the value 2.71828 (to five places of decimals), it follows that, except for the case where $t = 0$, $1/e^{t/T}$, or alternatively $e^{-t/T}$, will always be a positive fraction. Consequently also, and except for the case $t = 0$, the varying fraction $1 - (1/e^{t/T})$, or equivalently $1 - e^{-t/T}$, will always be a positive fraction.

In the case that $t = 0$, $e^{t/T}$ equals 1. In this case, $1/e^{t/T}$ will also be equal to 1, and the varying fraction $1 - (1/e^{t/T})$ will take the value $1 - 1 = 0$.

Next, because t denotes a period and so varies with time, it follows that the exponential function $e^{t/T}$, which we could equivalently write in a more familiar form as $e^{(1/T)t}$, is a function with a constant proportionate nominal rate of increase of $(1/T)$ per period. In other words, the reciprocal of the time constant, $1/T$, is the proportionate nominal rate of increase of the exponential function which is part of the varying fraction. (As we shall see, with continuous growth the effective rate of growth is greater than the nominal rate of growth because the growth, or 'compounding', is continuous, and this continuous nature of growth enhances the actual or 'effective' rate of growth over the nominal rate of growth.)

The common sense of this is that a small value of T means a fast proportionate nominal rate of increase $(1/T)$ in the exponential function, and consequently this implies a rapid rate of learning; it takes fewer periods for the plant to achieve 63.2 per cent of the transition from initial output to maximum output.

Consequently, the denominator, $e^{t/T}$, in the fraction, $1/e^{t/T}$, *increases* continuously through time. Conversely, of course, the value of the fraction $1/e^{t/T}$ itself *decreases* continuously through time.

In turn, and importantly, this means that the value of the varying fraction $1 - (1/e^{t/T})$ *increases* continuously through time.

Eventually, the value of $e^{t/T}$ becomes very large (with large values of t), and conversely the value of $1/e^{t/T}$ becomes very small and approaches zero. This means that the value of the varying fraction $1 - (1/e^{t/T})$ approaches a value of 1.

The preceding paragraphs may be illustrated by taking a value of $T = 2$ periods, to give the following numerical values:

Period, t	$e^{t/2}$	$1/e^{t/2}$	*Value of the varying fraction* $1 - (1/e^{t/2})$
0	1	1	0
1	1.64872	0.60653	0.39347
2	2.71828	0.36788	0.63212
3	4.48169	0.22313	0.77687
4	7.38906	0.13534	0.86466
5	12.18249	0.08208	0.91792
6	20.08554	0.04979	0.95021
.	.	.	.
.	.	.	.
.	.	.	.
12	403.42879	0.00248	0.99752

Using these illustrative values we may see the following eight things.

First, because the exponential function $e^{t/2}$ grows at a proportionate nominal rate of $1/2 = 0.5$ per period, the proportionate effective rate of growth for the exponential function is $e^{0.5} - 1 = 2.71828^{0.5} - 1 = 1.64872 - 1 = 0.64872$ per period.

Second, the denominator $e^{t/2}$ increases at a constant proportionate effective rate of 0.64872 from period to consecutive period. For example: 1.64872/1 = 1.64872; 2.71828/1.64872 = 1.64872; 4.48169/2.71828 = 1.64872; and so on.

Third and conversely, the fraction $1/e^{t/2}$ decreases at a constant proportionate rate from period to consecutive period. For example: 0.60653/1 = 0.60653; 0.36788/0.60653 = 0.60653; 0.22313/0.36788 = 0.60653; and so on.

Fourth, if a number (say, starting at 1) decreases at a constant proportionate rate, this means that the absolute amount of the decrease is bigger at first but that the absolute amount of the decrease becomes smaller as the number itself becomes smaller. For example: from $t = 0$ to $t = 1$, the absolute amount of the decrease in $1/e^{t/2}$ is from 1 to 0.60653, or 0.39347; from $t = 1$ to $t = 2$, the absolute amount of the decrease in $1/e^{t/2}$ is from 0.60653 to 0.36788, or 0.23865; from $t = 2$ to $t = 3$, the absolute amount of the decrease in $1/e^{t/2}$ is from 0.36788 to 0.22313, or 0.14475; and so on. Recapitulating, the absolute amounts of the successive decreases are: 0.39347, 0.23865, and 0.14475.

Fifth and conversely, to subtract such a number as $1/e^{t/2}$ which decreases in this way from any constant number (such as 1) means that the results of such subtractions will increase by exactly these (decreasing) absolute amounts.

To illustrate, the varying fraction $1 - (1/e^{t/2})$ increases in absolute amount as follows:

Period, t	*Value of the varying fraction* $1 - (1/e^{t/2})$	*Absolute amount of increase over preceding value*
0	0	–
1	0.39347	0.39347
2	0.63212	0.23865
3	0.77687	0.14475

Thus from one period to the next, the value of the varying fraction $1 - (1/e^{t/2})$, or more generally of $1 - (1/e^{t/T})$, increases by successively decreasing absolute amounts. The increase is big at first but progressively becomes smaller in absolute amount.

We are trying to describe a general pattern of the way in which improvement occurs under the learning effect. Taking periods of equal duration, it is a quite common pattern of improvement that, in any process which is being repeated, the improvement is quite marked, or 'big', in the first few periods (when the obvious difficulties caused by inexperience and/or lack of knowledge are being overcome), but the improvement then becomes less marked, or 'smaller', in succeeding periods as it becomes more difficult to improve on the level of proficiency attained. The formula of the varying fraction $1 - (1/e^{t/T})$ models this pattern of increase or 'improvement', as shown in the illustrative values for the absolute amount of the increase over the preceding value.

Sixth, eventually (as we can see for $t = 12$), the value of $e^{t/2}$ becomes very large; conversely, the value of $1/e^{t/2}$ becomes very small; and in turn the value of the varying fraction $1 - (1/e^{t/2})$ approaches a value of 1.

In turn this means that the time rate of production, according to the time constant progress function, approaches the maximum achievable. Note that the varying fraction, in approaching 1, can (mathematically) never go beyond that limiting value; in this model, it is not possible for improvement to go beyond whatever 'maximum' is achievable. And this is theoretically as it should be.

This pattern of increase in the varying fraction $1 - (1/e^{t/2})$, alternatively the varying fraction $1 - e^{-t/T}$, is the pattern of change which we require for describing the impact of the learning effect on the time rate of production.

Seventh, we can now explain how the value of the fraction 0.632 (63.2 per cent) arises. After the elapse of T periods, that is, when $t = T$ so that $t/T = 1$, we have:

$$\begin{aligned} 1 - e^{-t/T} &= 1 - e^{-1} \\ &= 1 - 2.71828^{-1} \\ &= 1 - 0.36788 \\ &= 0.63212 \end{aligned}$$

This value of the varying fraction will occur whenever $t = T$.

Eighth, we can explain how the value of the fraction 0.95 (95 per cent) arises. After the elapse of a number of periods equal to the time constant multiplied by 3, that is, when $t = 3T$ so that $t/T = 3$, we have:

$$\begin{aligned} 1 - e^{-t/T} &= 1 - e^{-3} \\ &= 1 - 2.71828^{-3} \end{aligned}$$

$$= 1 - 0.04979$$

$$= 0.95021$$

This value of the varying fraction will occur whenever $t = 3T$.

Table 8.4 shows values of the varying fraction for a selection of different possible values of the time constant T. By the nature of T as a parameter, only one numerical value of T can be taken for any given plant, so the column headings may also be taken as applying to plants (each of which has a learning pattern described by that particular value of the time constant).

We see that when $t = 0$, the value of the varying fraction is uniformly zero, regardless of the value of T.

We see that thereafter the value of the varying fraction varies positively with t (read down the columns of Table 8.4).

Figure 8.1 shows the pattern of variation. We can see how the varying fraction fits the pattern of rising quickly at first and then more slowly, according to the pattern of change just illustrated for $T = 2$. Thus the varying fraction follows the pattern of the top part of Figure 6.2, and this is exactly the pattern of improvement which we seek to model in the varying fraction.

Reading down any column of Table 8.4, we see that the varying fraction asymptotically approaches a limiting value of 1.

Reading across any row of Table 8.4, we see that the value of the varying fraction depends on T. The greater is T (and here we must be considering comparisons among different plants, each with a different parameter value for T), the smaller is the value of the varying fraction for any given value of t. That is, the greater the time constant, the less learning will have occurred by any given period t.

Finally, we see that whenever $t = T$ the value of the varying fraction is 0.63212, and whenever $t = 3T$ the value of the varying fraction is 0.95021.

We may now use the expression for the varying fraction in order to derive the formula for the time constant progress function as follows:

$$Q_t = Q_0 + (Q_{max} - Q_0) \begin{pmatrix} \text{Fraction varying} \\ \text{with } t \text{ and} \\ \text{dependent upon } T \end{pmatrix}$$

$$= Q_0 + (Q_{max} - Q_0)\,(1 - e^{-t/T})$$

Table 8.4

The varying fraction $(1 - e^{-t/T})$ for t and different values of the time constant T

Discrete time period, t	Values of $(1 - e^{-t/T})$ $T=1$	$T=2$	$T=3$	$T=4$	$T=5$	$T=6$	$T=7$	$T=8$
0	0	0	0	0	0	0	0	0
1	0.63212	0.39347	0.28347	0.22120	0.18127	0.15352	0.13312	0.11750
2	0.86466	0.63212	0.48658	0.39347	0.32968	0.28347	0.24852	0.22120
3	0.95021	0.77687	0.63212	0.52763	0.45119	0.39347	0.34856	0.31271
4	0.98168	0.86466	0.73640	0.63212	0.55067	0.48658	0.43528	0.39347
5	0.99326	0.91792	0.81112	0.71350	0.63212	0.56540	0.51046	0.46474
6	0.99752	0.95021	0.86466	0.77687	0.69881	0.63212	0.57563	0.52763
7	0.99909	0.96980	0.90303	0.82623	0.75340	0.68860	0.63212	0.58314
8	0.99966	0.98168	0.93052	0.86466	0.79810	0.73640	0.68109	0.63212
9	0.99988	0.98889	0.95021	0.89460	0.83470	0.77687	0.72355	0.67535
10	0.99995	0.99326	0.96433	0.91792	0.86466	0.81112	0.76035	0.71350
11	0.99998	0.99591	0.97444	0.93607	0.88920	0.84012	0.79225	0.74716
12	0.99999	0.99752	0.98168	0.95021	0.90928	0.86466	0.81991	0.77687

Source: see text.

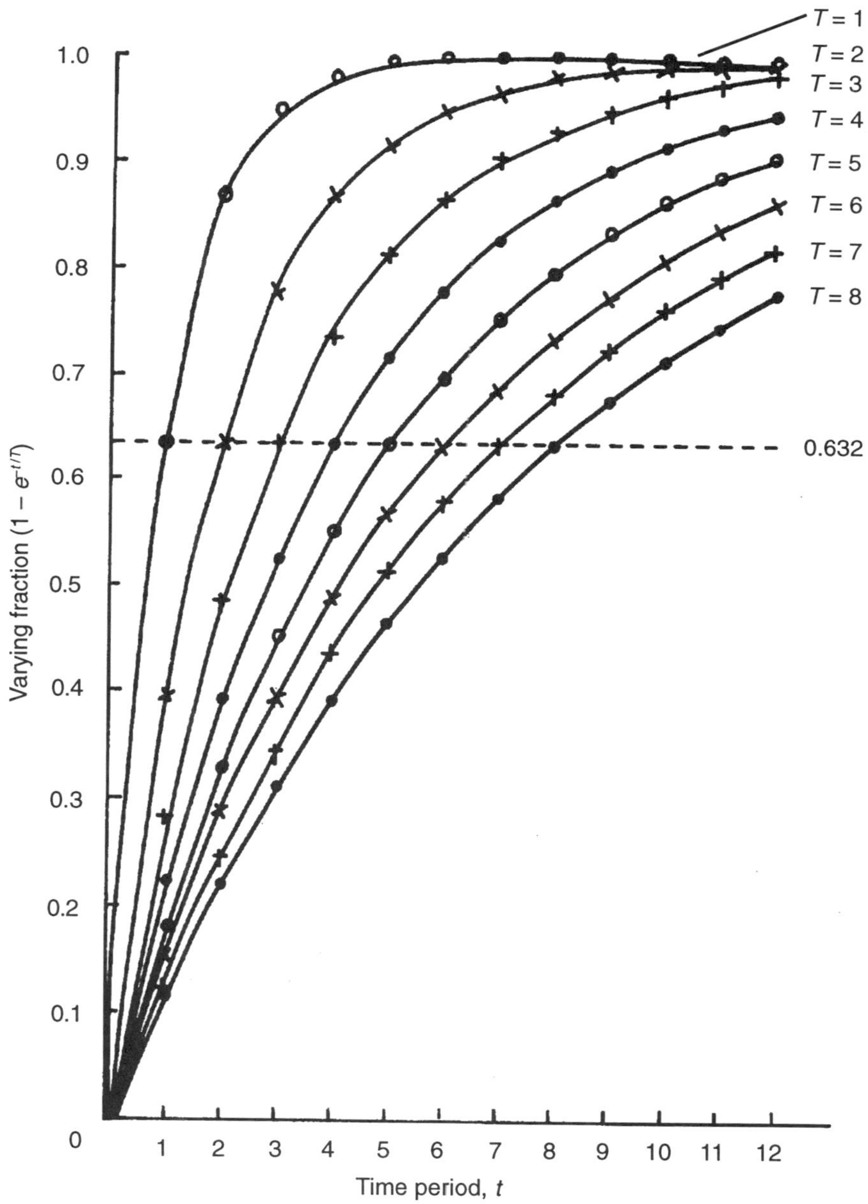

Figure 8.1 The varying fraction of the time constant progress function

The last line of the equation gives the formula for the time constant progress function.

If you read further in the literature on the time constant progress function, the term $(Q_{max} - Q_0)$ may be denoted by a single term which is referred to either as the 'transient part of the output rate' or as the 'dynamic gain'. The ratio $(Q_{max} - Q_0)/Q_0$ is referred to as the 'normalized dynamic gain', and is obviously useful to a production manager for its information on the relative extent of the eventual improvement to be expected in the time rate of production. (D.R. Towill, 'The Use of Learning Curve Models for Prediction of Batch Production Performance', *International Journal of Operations and Production Management*, vol. 5, no. 2, 1985, pp.16–17, gives a sample of eight studies which show an average normalized dynamic gain of 1.41, or 141 per cent. In other words, on average in this sample $Q_{max}/Q_0 = 2.41$, or Q_{max} was more than double Q_0. The eight time constants from these studies range from 24 weeks for a large work group engaged in packaging in a pharmaceuticals plant, through 20 weeks for a medium-sized work group engaged in the start-up of a rolling mill in the steel industry, to 3 weeks for one worker engaged in electrical switch assembly. This shows how the time constant differs according to type of process, and how, for example, it would only be meaningful to compare, say, the time constants among steel industry rolling mills if these were to differ.)

Estimating the values of the parameters in the time constant progress function requires the use of iterative non-linear estimation methods which are too complex to be explained here. Instead, we continue with an illustration (based on simple parameter estimates) to show how the time constant progress function fits data on the time rate of production when the plant is subject to the learning effect.

To do this, we return to the data of Table 6.1 on the assembling of propelling mechanisms. The relevant figures are given in Table 8.5.

We take as an estimate of Q_0 the rate of output in the initial week ($t = 0$) of 9,300 mechanisms. We next take the rate of output during the twelfth week ($t = 11$), 17,000 mechanisms, as an estimate of the eventual maximum rate of output period, Q_{max}.

Finally, we observe that during the third week of production (when $t = 2$; that is, after the elapse of two periods (weeks)) the rate of output achieved, 14,600, gives about 68.8 per cent of the difference between the initial rate of output and the eventual maximum rate of output: (14,600 – 9,300)/(17,000 – 9,300) = 0.688. The time constant is that number of periods which have to elapse before 63.2 per cent of the transition between initial output and eventual

Table 8.5
Illustration of the time constant progress function applied to assembling of mechanical pencil mechanisms

	Time constant ($T = 2$) progress function						Estimated	Actual	Percentage
Week, t	Q_0	+	$(Q_{max} - Q_0)$	×	$(1 - 2.71828^{-t/2})$	=	$\hat{Q}_t$	Q_t	error
0	9,300	+	(17,000 – 9,300)	×	$(1 - 2.71828^{-0/2})$	=	9,300	9,300	0
1	9,300	+	(17,000 – 9,300)	×	$(1 - 2.71828^{-1/2})$	=	12,330	11,900	+3.6
2	9,300	+	(17,000 – 9,300)	×	$(1 - 2.71828^{-2/2})$	=	14,167	14,600	–3.0
3	9,300	+	(17,000 – 9,300)	×	$(1 - 2.71828^{-3/2})$	=	15,282	15,300	–0.1
4	9,300	+	(17,000 – 9,300)	×	$(1 - 2.71828^{-4/2})$	=	15,958	15,800	+1.0
5	9,300	+	(17,000 – 9,300)	×	$(1 - 2.71828^{-5/2})$	–	16,368	15,900	+2.9
6	9,300	+	(17,000 – 9,300)	×	$(1 - 2.71828^{-6/2})$	=	16,617	16,200	+2.6
7	9,300	+	(17,000 – 9,300)	×	$(1 - 2.71828^{-7/2})$	=	16,767	16,000	+4.8
8	9,300	+	(17,000 – 9,300)	×	$(1 - 2.71828^{-8/2})$	=	16,859	16,300	+3.4
9	9,300	+	(17,000 – 9,300)	×	$(1 - 2.71828^{-9/2})$	=	16,914	16,400	+3.1
10	9,300	+	(17,000 – 9,300)	×	$(1 - 2.71828^{-10/2})$	=	16,948	16,000	+5.9
11	9,300	+	(17,000 – 9,300)	×	$(1 - 2.71828^{-11/2})$	=	16,969	17,000	–0.2

Source: see text and Table 6.1.

maximum output is achieved, and 68.8 per cent is reasonably close to 63.2 per cent, at least for the present purposes of illustrating and explaining the time constant progress function. Accordingly, we may characterize the learning pattern for this worker assembling propelling pencil mechanisms by the time constant $T = 2$ weeks as an estimate of this key parameter.

Using these estimated values of the three parameters: $Q_0 = 9{,}300$ per week; $Q_{max} = 17{,}000$ per week; and $T = 2$ weeks, we may work out the estimated values of output per period, $\hat{Q}_t$, according to the time constant progress function.

In each row the expression for the varying fraction is shown in full.

For the initial week, $t = 0$, the value of the varying fraction is 0, so $Q_t = Q_0 = 9{,}300$.

For the second week, $t = 1$, $\hat{Q}_1 = 12{,}330$, against an observed Q_1 of 11,900, so the difference between the value of output according to the time constant progress function and the observed value is 3.6 per cent, a negligible difference.

For the third week, $t = 2$, $\hat{Q}_2 = 14{,}167$, against an observed Q_2 of 14,600, so the difference between the value of output according to the time constant progress function and the observed value is –3 per cent, a negligible difference.

Continuing in Table 8.5, in most cases the percentage difference between the value of weekly output according to the time constant progress function and the observed value is negligible. This shows that the model of the time constant progress function provides a good representation of the actual observations.

We may illustrate how the time constant progress function fits the pattern shown by the actual observations in Figure 8.2. In Figure 8.2, dots represent the actual observations, while crosses represent the values estimated according to the time constant progress function. The time constant progress function is drawn as a curve through the estimated values.

We can see that the curve of the time constant progress function provides a reasonable representation of the observed values, and certainly fits the pattern of rising quickly at first and then approaching some maximum value.

We must now consider the impact of all this on the annual rate of profit in a given plant subject to the learning effect.

8.3 Learning and the efficiency rate of profit

The efficiency rate of profit is measured as the annual flow of profit divided by the acquisition cost of the capital stock installed. If we wish to analyse the

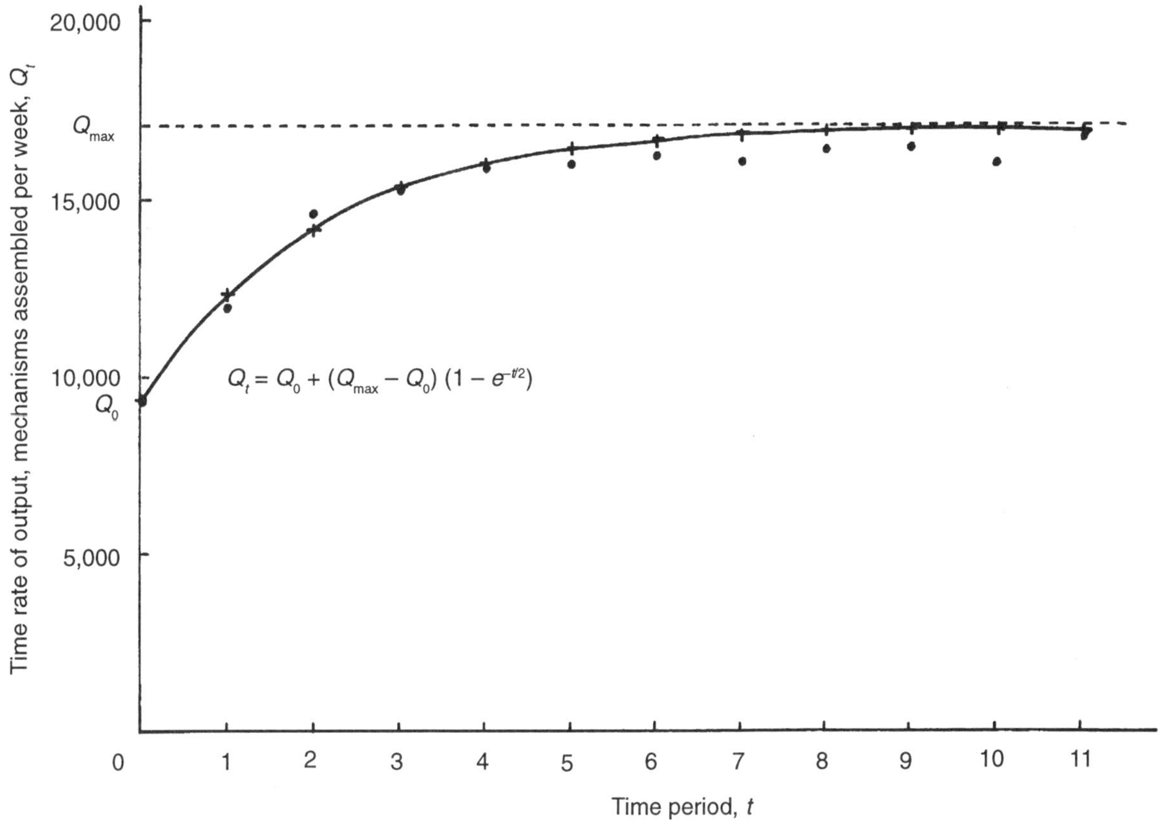

Figure 8.2 Time constant progress function for the assembling of mechanical pencil mechanisms

impact of the learning effect on the rate of profit, we have to look at what happens to the various relevant variables during each year. Note that because the rate of profit is usually taken as a rate *per annum* we now switch to a discussion specifically in terms of years rather than more generally in terms of periods. Consequently, the series t must now be taken to denote years. Consequently also, the time constant parameter, T, must be given in a number of years elapsing. However, the following argument is also more generally applicable to periods of any duration (specifically, those periods (shorter than one year) which may be more appropriate to the time-scale in which production and learning occur and which are therefore better suited to measuring the time constant). Inescapably, any analysis for periods shorter than one year suffers the drawback that each profit flow must initially be calculated for this shorter period (also altering p_K to a rate per the shorter period), and the subsequent aggregating of the various profit flows into an annual profit flow considerably complicates the analysis. Consequently, we proceed to the simpler analysis using periods of one year.

The progress of the argument to follow is summarized in Figure 8.3.

The annual flow of profit, Π_t, can be written as the profit flow equation (see Chapter 1), where we subscript the relevant variables with regard to t where appropriate, but we shall assume a constant unit materials cost, denoted *UMC*:

$$\Pi_t \quad = \quad [p_Q - (UMC + p_L\,(L_t/Q_t) + p_K\,(K_t/Q_t))]\;Q_t$$

For a given plant, we assume a constant annual labour input and a constant amount of capital installed, so L_t and K_t, although subscripted with regard to t, are assumed to remain constant. Product price, p_Q, is assumed constant.

The expression in square brackets is the unit net margin. For the calculation of annual profit, the unit net margin must be defined with respect to *year* average unit labour requirement, L_t/Q_t, and year average unit capital requirement, K_t/Q_t.

The learning effect may be analysed through the learning curve. The particular item unit labour requirement learning curve leads to a decreasing unit labour requirement for each item. Assuming that the quantity of labour input per year remains constant and also that the quantity of capital installed remains constant, then the annual rate of production increases from one year to the next, and this pattern of increase can be described and modelled by the time constant progress function. Because of the increase in the time rate of production, with constant labour and capital input, the year average unit labour

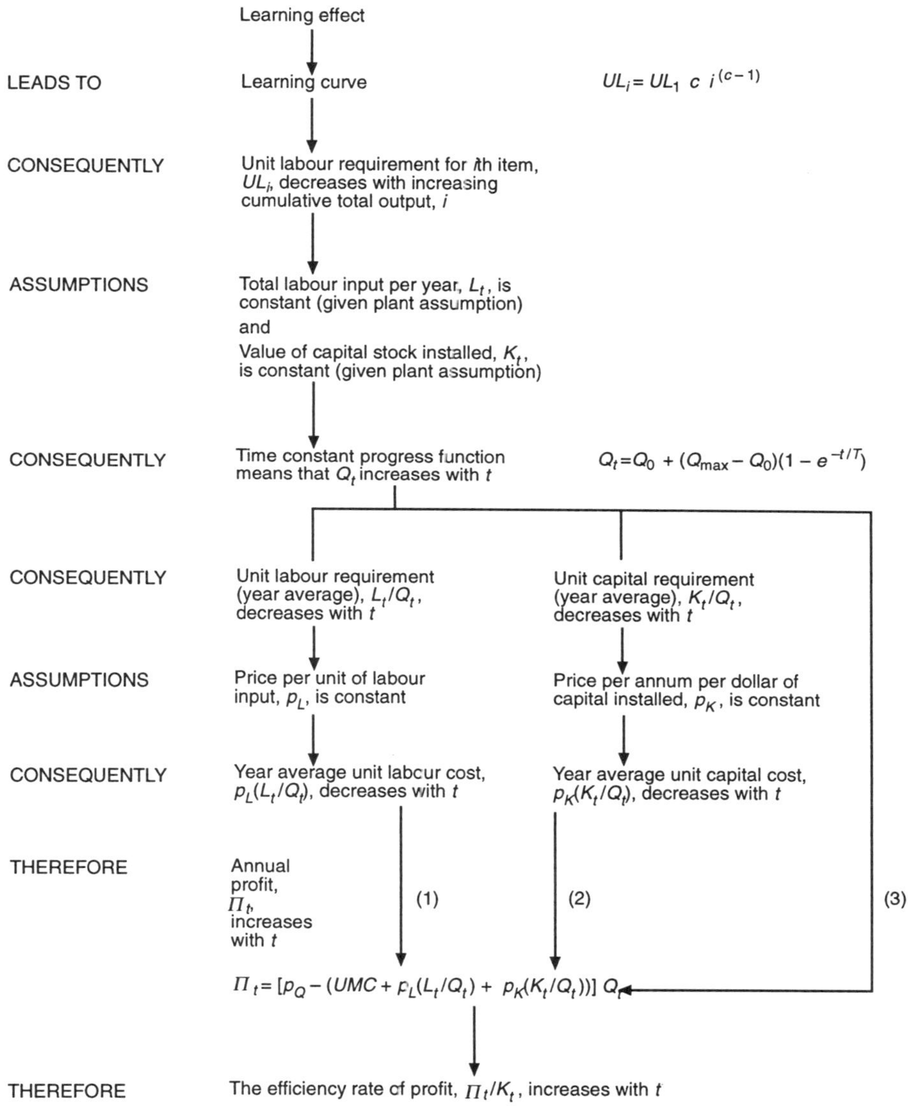

Figure 8.3 The learning curve, the time constant progress function, and the efficiency rate of profit

requirement and the year average unit capital requirement, each decrease with the increase in Q_t.

Assuming that the price of labour, p_L, and the price of capital, p_K, each remain constant, these decreases in year average unit input requirements lead to decreasing unit factor input costs.

Consequently, the year average unit net margin increases.

At the same time that the unit net margin is increasing, the annual rate of production is also increasing, and together the two result in an increase in annual profit.

We need to illustrate in a graph the way in which unit labour cost, averaged over the year, changes with *t*. To do so we may use a simple arithmetical example. The expression for unit labour cost may be written as follows:

Year average unit labour cost	=	Price per unit of labour	×	Year average unit labour requirement
	=	$p_L(L_t/Q_t)$		

Substituting the time constant progress function for Q_t, we get:

$$\text{Year average unit labour cost} = p_L \frac{L_t}{Q_0 + (Q_{max} - Q_0)(1 - e^{-t/T})}$$

We are assuming that p_L and L_t are constants, and in the time constant progress function Q_0 and Q_{max} are also constants. To make the example as simple as possible, we may assume that p_L, L_t, and Q_0 all take the value 1.

Furthermore, we assume that Q_{max} takes the value 2 (this amounts to assuming that the maximum time rate of output will be double the initial rate of output). The value of 2 for Q_{max} is convenient because then $(Q_{max} - Q_0) = 1$. (These arithmetical assumptions are merely for the convenience of illustration; we would get the same *pattern* of results if we took 'real' numbers, but the arithmetic would be more complicated.)

With these assumed values we may write the numerical expression for year average unit labour cost, with the initial value equal to 1, as:

$$\text{Year average unit labour cost} = 1 \times \frac{1}{1 + 1(1 - e^{-t/T})}$$

Table 8.6
Hypothetical illustration of the trend in year average unit labour cost under the learning effect

	$T = 2$	
Year, t	Annual output, $1 + 1(1 - e^{-t/T})$	Year average unit labour cost, $1 \times \dfrac{1}{1 + 1(1 - e^{-t/T})}$
0	1	1
1	1.39347	0.71763
2	1.63212	0.61270
3	1.77687	0.56279
4	1.86466	0.53629
5	1.91792	0.52140
6	1.95021	0.51277
7	1.96980	0.50767
8	1.98168	0.50462
9	1.98889	0.50279
10	1.99326	0.50169
11	1.99591	0.50102
12	1.99752	0.50062

Source: see text.

Suppose that to continue the illustration we choose $T = 2$ years, as in the assembling of propelling pencil mechanisms we took $T = 2$ weeks. Table 8.6 illustrates what happens to year average unit labour cost.

The first column of Table 8.6 gives the values of the time constant progress function. These are the values of Q_t. We see the familiar pattern that Q_t increases quickly at first and then more slowly, and asymptotically approaches the maximum value of 2.

The last column of Table 8.6 gives the values of $1 \times (1/(1 + 1(1 - e^{-t/2})))$. With the assumed values, these are not only the values of $1/Q_t$ but are also the values of the year average unit labour cost.

We can see that this year average unit labour cost decreases quickly at first, and then more slowly, and asymptotically approaches a limiting value of 0.5. In other words, as the annual rate of production asymptotically approaches double its initial rate, year average unit labour cost asymptotically approaches half its initial value.

Figure 8.4 shows a graph of this year average unit labour cost, showing

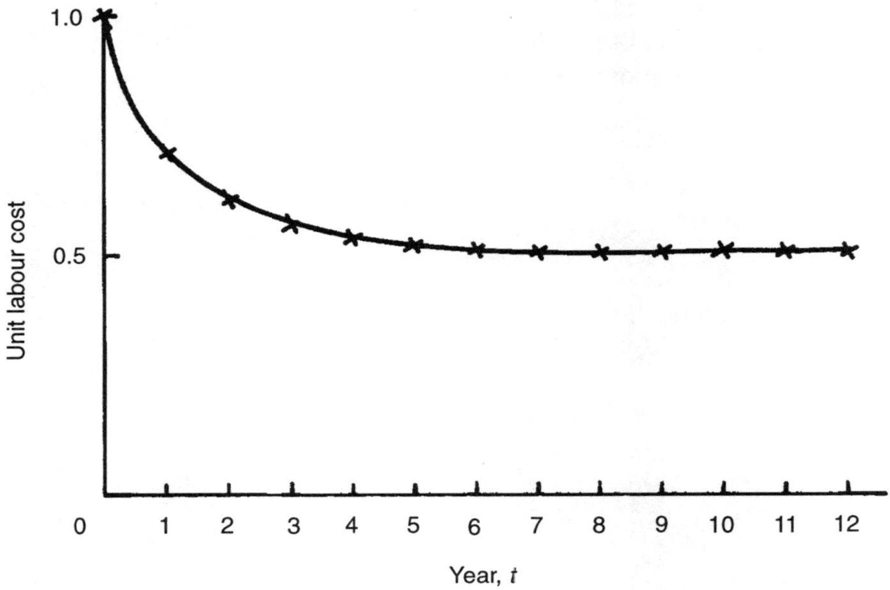

Figure 8.4 Trend in unit labour cost (year average) under the learning effect

how year average unit labour cost varies as a function of elapsed time, t. We can see clearly how year average unit labour cost decreases quickly at first and then more slowly. This time series trend comes about because, under the learning effect, the particular item unit labour requirement decreases with cumulative total output and so affects the annual rate of output in the way modelled by the time constant progress function.

Note that Figure 8.4 is very similar to the bottom part of Figure 6.2. Figure 8.4 models (for periods of years rather than weeks) the trend of unit labour cost which could be obtained by multiplying Figure 6.2's unit labour requirement by a constant price of labour.

Returning to Figure 8.3, the arrow labelled (1) shows that year average unit labour cost decreases with decreasing unit labour requirement.

The argument for the trend in unit capital cost is exactly parallel to that for the trend in unit labour cost. The quantity of capital installed and the rate of depreciation are both assumed constant so that the annual flow of depreciation is constant, and unit capital cost is annual depreciation divided by the annual flow of output, Q_t, so that when Q_t increases under the impact of the learning effect (in the way modelled by the time constant progress function), unit capital cost decreases correspondingly. This is shown in Figure 8.3 by the arrow labelled (2).

At this point the assumption of a constant unit materials cost, *UMC*, should be recalled. This is a simplifying assumption, and we have seen that unit material requirement may itself be subject to a learning effect. But our main concern is with unit labour requirement, and assuming a constant unit materials cost means that the subsequent analysis understates the impact on profit, and so can be regarded as a cautious rather than an exaggerated analysis.

The assumption of constant unit materials cost, together with decreasing year average unit labour cost and decreasing year average unit capital cost, means that, with a constant price of output, the year average unit net margin will increase from one year to the next.

Figure 8.5 illustrates all this. (Figures 8.5 and 8.6 are based on: unit labour cost, Table 8.6; unit capital cost, one-half unit labour cost; unit materials cost, a constant value of 2; product price equal to unit total cost of 2.844185 in $t =$ 3; time rate of production, Table 8.6.) The diagram shows the years, t, marked along the horizontal axis and (illustrative) dollar values on the vertical axis. The curve labelled 'Unit labour cost (year average)' reproduces Figure 8.4. Unit capital cost follows a trend similar to that of unit labour cost, and this is indicated accordingly in another curve.

Unit labour and capital cost combined is simply the vertical sum of these two curves and is accordingly drawn as such.

Unit materials cost is assumed constant in each year. Consequently, unit total cost (averaged over the year) is simply this constant added to unit labour and capital cost combined. In other words, the unit labour and capital cost curve is shifted upwards by the constant amount of unit materials cost.

Accordingly, year average unit total cost is drawn as a curve in Figure 8.5. We can see how unit total cost decreases with t, the decrease being rapid at first and then becoming more gradual.

Unit total cost may be taken as a measure of performance, and in Figure 8.5 this measure of performance is given as a function of elapsed time. Accordingly, the unit total cost curve may be identified as a progress function. This trend of unit total cost decreasing with elapsed time is a very important trend for plant managers to understand.

It is assumed the price of output, p_Q, or 'product price', remains constant. Consequently, we may represent p_Q as a horizontal line in Figure 8.5.

Obviously, it is possible for product price to be above unit total cost from

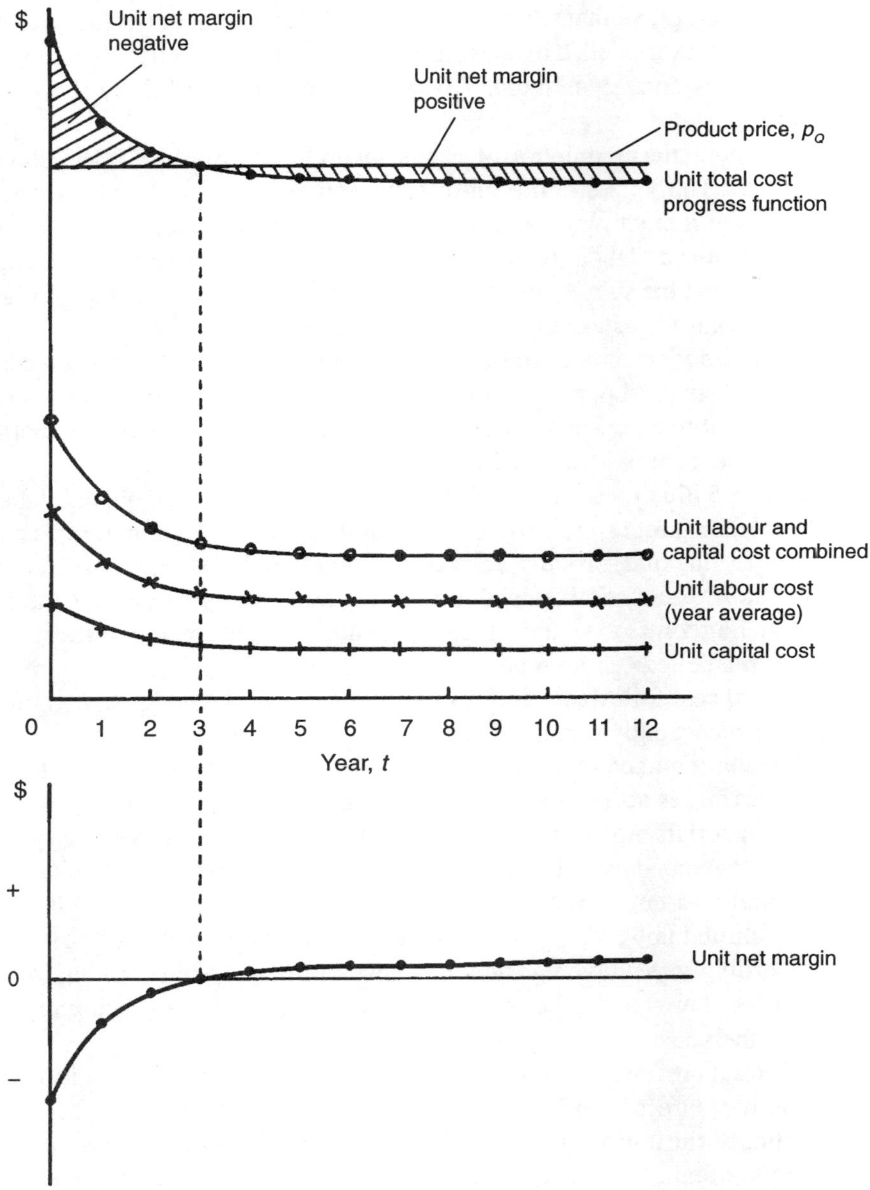

Figure 8.5 Trend in the unit net margin under the learning effect

the start, so that profits are made from the start and continue to be made as well as to increase. This situation is the most common one, and it causes no problems for managers.

It is also possible for product price to be, and to remain, below unit total cost, in which case the product generates only losses, although the annual amount of the loss will decrease.

Figure 8.5 depicts an in-between situation, which is the interesting one to analyse. The product (plant) makes a loss during each of years 0 to 2 inclusive, breaks even during year 3, and then makes a profit during all years from, and including, year 4.

When we represent p_Q as a horizontal line in the position chosen, we can see immediately what is happening. During years 0, 1, and 2 the year average unit total cost is above, or greater than, product price, so that the unit net margin (averaged over the year) is negative, as indicated by the shading in the diagram.

We can also see that the (negative) gap between unit total cost and product price is very considerable to start with, but (in absolute value) decreases quite rapidly with each of the successive years, as the product (plant) proceeds along its progress function.

During year 3, year average unit total cost is equal to product price, so that the year average unit net margin is zero. In other words, during year 3, the plant exactly covers the annual costs of production from annual sales receipts.

During year 4, and during all succeeding years, year average unit total cost is below, or less than, product price, so that the year average unit net margin is positive, as indicated by the shading in the diagram. Moreover, the unit net margin increases with the years although this increase becomes more gradual over time.

The bottom part of Figure 8.5 (vertically aligned with the top part) shows more directly what is happening to the year average unit net margin. The vertical axis here measures the vertical distance between the horizontal line representing p_Q and the unit total cost progress function. We can then see that year average unit net margin is negative during years 0, 1, and 2; has a zero value during year 3; and has a positive value in year 4 and all succeeding years. We can also see the pattern of the increase in the year average unit net margin: rapidly at first and then more slowly.

All this comes about because of the impact of the learning effect.

The year average unit net margin is one component in the equation for the annual flow of net profit. The other component is the time rate of output per annum, Q_t. We now consider the behaviour of Q_t under the learning effect.

In a given plant subject to the learning effect, so that the particular item unit labour requirement decreases according to the learning curve, and where the plant's quantity of the annual labour input (measured, say, in labour-hours) is constant, the time rate of production will follow the trend described by the time constant progress function. Thus in Figure 8.3 we can show by the arrow (3) that Q_t increases with t according to the time constant progress function.

How does this affect the annual flow of profit? The trend in annual profit is illustrated in Figure 8.6 (where all parts are vertically aligned).

The top part of Figure 8.6 reproduces the year average unit net margin from Figure 8.5. The middle part of Figure 8.6 shows the time constant progress function for annual output of $1 + 1(1 - e^{-t/2})$; see Table 8.6.

The annual flow of profit is then the product of the vertical distances of each of these from its respective horizontal axis.

Accordingly, the bottom part of Figure 8.6 has a curve depicting the annual profit. In each of years 0, 1, and 2, annual profit is negative, because the year average unit net margin is negative. In other words, the product (plant) makes a loss during each of these years. However, the amount of the annual loss decreases with each succeeding year.

During year 3, the total annual profit is zero, because the unit net margin is zero. The product (plant) exactly covers its annual costs of production from its annual sales receipts.

During year 4, and each succeeding year, annual profit is positive. Moreover, because *both* the year average unit net margin is increasing *and also* the time rate of (annual) production is increasing, the increase in annual profit, from one year to the next, continues to be a noticeable increase. In other words, the increase in the year average unit net margin and the increase in the time rate of production reinforce each other, and produce *a more marked increase*, from one year to the next, in the annual flow of profit.

The annual flow of profit receives, over time, a double boost: one boost from the increase in the unit net margin; and another boost from the increase in the annual quantity of output produced. Each of these is a result of the learning effect.

Finally, and straightforwardly, because we are assuming that a given plant operates with a constant value of capital installed, the efficiency rate of profit per annum, computed as the ratio between the annual flow of profit and the value of capital installed, follows proportionately exactly the same trend as the annual flow of profit.

Accordingly, in the situation which has been depicted, the plant will have a quite substantial rate of loss in the first year or so, but the rate of loss will

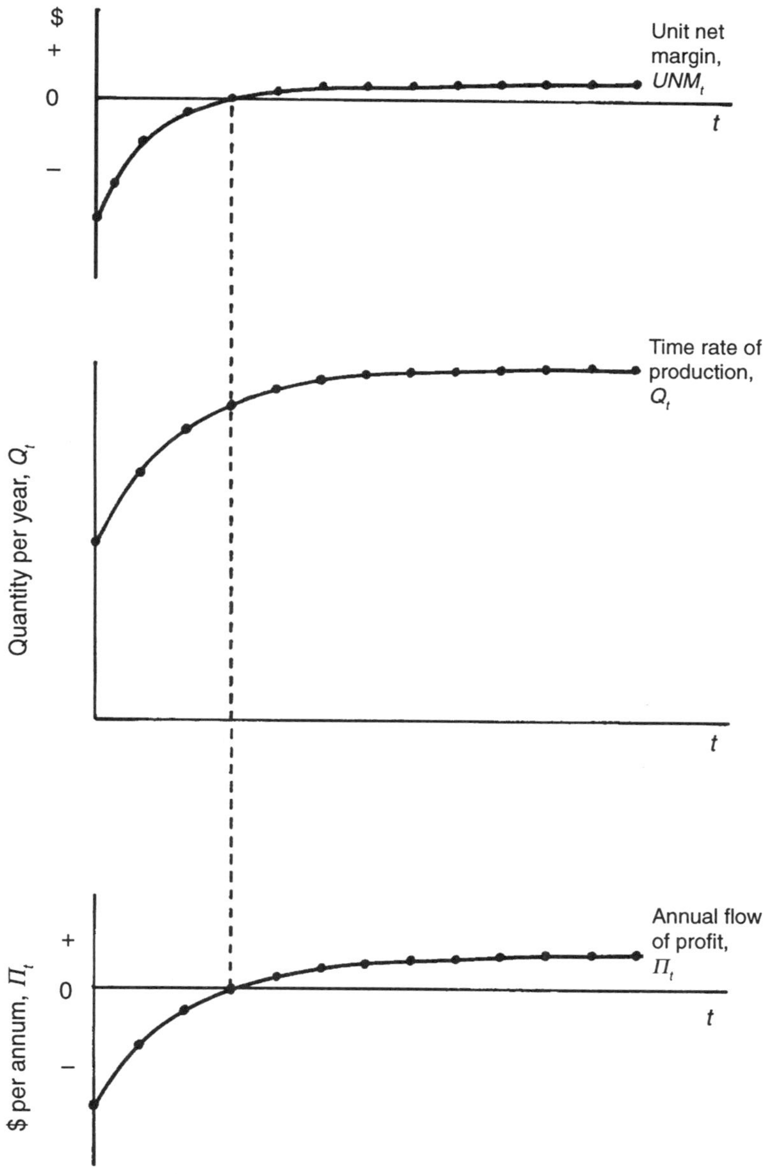

Figure 8.6 Trend in profit under the learning effect

diminish quite quickly. The plant will then break even, and will thereafter show an increasing efficiency rate of profit.

Obviously, the initial periods (years) during each of which a (quite substantial) loss may be made are likely to create problems, and these problems have been studied, particularly by production engineers, as the so-called 'start-up' problems.

Any enterprise which faces a learning situation in one of its plants should be prepared to finance losses in that plant during the first few periods (years). The managers of the plant should also look to ways (such as training, work study, and incentive schemes) of speeding up the learning effect and so reducing the number of periods (years) during which losses have to be sustained.

If the enterprise has a 'portfolio' of products, some of which are established and are well along their progress functions and so are generating profits, such profits can be used to finance the losses on the new products during the interim before such new products themselves begin to generate profits.

These issues are dealt with in management textbooks under the heading of the 'product portfolio matrix', the gist of which is to use profits (more strictly, a financial surplus or 'internal finance') generated by well-established profitable 'cash-cow' products to finance: (a) research and development for new products; (b) the installation of fixed capital to produce such products; and (c) the initial loss-making periods (years) when such new products are launched on the market and are making their way along the progress function (or 'down the learning curve') towards ultimate profitability and 'cash-cow' status.

The learning effect and its impact on profit through the learning curve and the time constant progress function are thus important for strategic management.

Index

DATE DUE